THE MISMAPPING OF AMERICA

The Mismapping of America

Seymour I. Schwartz

The University of Rochester Press

First published 2003

The University of Rochester Press
668 Mount Hope Avenue
Rochester, NY 14620, USA

and at Boydell & Brewer, Ltd.
P.O. Box 9
Woodbridge, Suffolk 1P12 3DF
United Kingdom

ISBN 1-58046-129-8

Library of Congress Cataloging-in-Publication Data

Schwartz, Seymour I., 1928-
The mismapping of America / Seymour I. Schwartz.
p.cm.
Includes bibliographical references and index.
ISBN 1-58046-129-8 (alk. paper)
1. Cartography—United States—History. 2. United States—History—Errors, inventions, etc. I. Title.

GA405 .S32 2003
912.73—dc21

2002155952

British Library Cataloging-in-Publication Data
A catalogue record for this book
is available from the British Library.

Designed and typeset by Christine Menendez
Printed in the United States of America
This publication is printed is on acid-free paper.

To Richard, Kenneth, and David
for their constant support

Contents

ILLUSTRATIONS

Preface

It has often been said that a picture is worth a thousand words. This is true for a portrait, a landscape, the rendition of a structure, or the expression of the artist's mood. Even more salient among the circumstances in which a graphic depiction offers significantly more than a verbal description is the representation of geography—a map! A map provides an imprint of the discovery of a new land mass, the topography of that land, a defined coastline, the course of a river, a mountain range, a boundary line, or the location of a specific locale, in addition to offering a sense of relative distances between the items incorporated on the pictorial document.

Maps simplify and facilitate the dissemination of geographic knowledge. They can convey more factual information, more nuance, or more texture than a multitude of words. And the information is often presented in a such an attractive fashion, replete with real or fanciful regional fauna and flora, portraits, allegories, and decorative cartouches, that maps are truly works of art. Maps educate and titillate!

Geographic knowledge, as is true for all sciences, is the product of observation and description. An area is viewed and a picture is imbedded in the brain. Then an attempt might be made to communicate the newly acquired knowledge to others by describing it using verbal or visual statements, or a combination of the two. Before the advent of the European printing press in the middle of the fifteenth century (a form of printing was used in China as early as the eleventh century), scribes produced textual manuscripts and maps on papyrus, parchment, or vellum.

The earliest medieval maps were drawn from narrative sources, and it was not until the fourteenth or fifteenth century that maps included information gathered from exploration and experience. Even that information was subject to individual interpretations, errors, and divergences on the part of the copyists, and, therefore, reliability was compromised. The receipt of disseminated information was initially limited to the clergy, the wealthy, and a few intellectuals. The application of movable type printing and the printing press to texts and maps in the mid-fifteenth century had a profound effect in that it expedited the spread of geographic facts, achieved uniformity, and extended the life of verbal descriptions and pictorial representations of newly acquired as well as established knowledge.

Verbal and visual statements, appearing in print, spread information, at first to inquiring academicians, later to those who were dependent on the imparted knowledge for their vocation, including those dedicated to discovery, exploration, and settlement. Eventually, the facts spread more widely to the inquisitive public at large.

But occasionally, in the process of expanding the intellectual horizons, errors in interpretation resulted in verbal or graphic misrepresentations. Some of the errors became ingrained permanently, or were perpetuated for prolonged periods of time. Maps, as powerful instruments for the dissemination of information, have also spread and nurtured misinformation. At times, maps have depicted fancies rather than facts; at other times, they have shown wishes rather than wisdom; at still other times, the illustrated errors were based on *alleged* facts.

Some of the errors that appear on maps have played significant roles in the history of the North American continent and the United States of America. In a time span that extended from the discovery of the continent by the European powers to the establishment of the new nation, five meaningful errors have had significant historic consequences. Perhaps the most blatant error is perpetuated in the name of that continent and the nation's name itself. The continent and nation bear the name of a man who not only never set foot on the North American continent, but played neither a primary nor a significant role in the discovery of the contiguous South American continent. What was the sequence of events that led to the imprinting of that name on the two newly discovered continents in the Western Hemisphere, and how did the naming achieve permanence?

In what way was the sixteenth-century definition of the East Coast of the future United States of America compromised by a major error of interpretation, perhaps based on wishful thinking, by an adventuresome explorer? That misinterpretation, which had great implication regarding potential commercial traffic between western Europe and the Orient, can be ascribed readily to a logical conclusion.

The Northwest Passage is a geographic term that is associated with its own transitions. Stemming, in the latter half of the sixteenth century, from the commercial desires of England—a country that, at the time, had evolved as the dominant naval power—the history of the Northwest Passage is replete with adventure, romance, tragedy, and accomplishment. In the processes of exploration and the search for those who gave their lives while exploring the area, what was originally a simplistic misrepresentation became modified to the point of defining the proof of its actual existence. Over the almost four centuries between the myth and the fact, maps depicted the series of dramatic events that actually led to the discovery of the passage.

How did the improbable sequential activities of a mythical Amazonian queen, a famous conquistador, a distinguished English mathematician, a Jesuit priest, and a Spanish king participate in the pendulum swing of the perception of California, first, as a peninsula, then, as an island, and, ultimately, as a peninsular part of the North American continent?

And, as ultimate evidence of the role of cartographic errors in the history of the United States of America, the map that was used by all parties at the Treaty of Paris in 1783 to define the boundaries of the new nation included a small apocryphal island that served as a specific reference point.

Surely, the words of the English adventurer John Smith, who gained fame by his exploits in Virginia, and who mapped Virginia, Chesapeake Bay, and New England, continue to ring true. In his 1624 *The Generall Historie of Virginia, New-England, and the Summer Isles*, Smith stated: "As Geography without History seemeth as a carkasse without motion, so History without Geography wandereth as vagrant without a certaine habitation."

But, during a time span that extended from the discoveries made by Christopher Columbus to the execution of a treaty that established the boundaries of the new nation of the United States of America—a period of almost three hundred years—five major errors, which were expressed on maps, had significant historic influences. America has been misnamed, misrepresented, and mismapped. Many people have died because of these mistakes.

Many others died trying to correct them. And the most significant of these errors will never be corrected!

Acknowledgments

I am grateful to Melinda Beard, Martin Greenberg, and Louise Goldberg for editorial assistance that produced significant improvements. I thank Martha Smith and Michael Malerk for their photographic expertise. Louis Cardinal of the National Archives of Canada provided great assistance in the acquisition of important images. And I am thankful to Timothy Madigan and Molly Cort of the University of Rochester Press for their cooperation and input.

CHAPTER ONE

The Greatest Misnomer on Planet Earth

Almost five hundred years after the print dried on the document, America's birth certificate now has come to rest in the Western Hemisphere. The assigned name of "America" is dramatically displayed on an ornate pictorial representation of a world that had been recently updated to include a landmass in the Western Hemisphere, separating Europe from Asia. The picture, a *mappa mundi* or world map, drawn and printed in what is now France, had been housed in the sixteenth-century castle of Prince Johannes Waldburg-Wolfegg in southern Germany for at least two hundred fifty years. It is the only known extant copy. After nearly a decade of negotiations with the German government, the owner was permitted to offer the map for sale to the United States. In 2001, it was purchased for the princely sum of ten million dollars, and now highlights the largest repository of maps in the world, the Geography and Map Division of the Library of Congress, with its 4.5 million maps that graphically chronicle the evolution of the United States from its discovery to the present (see figure 4).

The word "America" evokes a sense of power, patriotism, and pride. It conjures up images of an expansive landmass replete with natural beauty and fertile fields. It connotes financial and commercial success. It stands for hope, freedom, and democracy—the "American way." The United States of America is characterized by a uniquely diversified population, emblematic foods, and idiomatic speech. "American" is the adjective attached to a dream. But all these positive images incorporate a name that stands for a lie!

Stephen Vincent Benet wrote, "I love American names." The names assigned to locations on the North American continent are unequaled in the diversity of their origin as a consequence of the diversity of the people residing on the land. The pieces of the mosaic that constitute the United States of America, and the locales within those pieces, have been anointed with names drawn from the history of the evolution of the nation. To the original descriptive terms imposed by the Native Americans, names have been added from essentially all of Europe, as well as from Asia, Polynesia, and biblical sites during the last five centuries. The names found on the continent today memorialize saints, kings, aristocrats, heroes, settlers, developers, and entrepreneurs.

Perhaps the most curious, most historically enigmatic, and, many have argued, most inappropriate name to appear on maps depicting the northern continent in the Western Hemisphere is "America" itself. Several questions arise. Who was the man from whom the name derives? Why was he so honored? What was the sequence of events that led to the imposition and perpetuation of that honor? What other names could have or should have been considered?

The planet Earth, on which we reside, derives its name from the ancient Teutonic word for "land." The larger continuous land masses on our planet are referred to as continents. Of the seven continents, two of the three continents known to European civilized man before the sixteenth century, the so-called Ptolemaic World, were thought to have been given the names of women. Europe took its name from Europa, the daughter of a Phoenician king. Mythology relates that all-powerful Zeus, King of the Gods, was so impressed with her beauty that he assumed the appearance of a white bull, and abducted her to Crete where she bore him three sons, including Minos, who became one of the three judges in the Underworld after his death.

Asia was also thought by some to have been named for a woman. In Greek mythology, Asia was the wife of the Titan Iapetus and the mother of Prometheus, the Titan who stole fire from the gods and was punished by being tied on a mountain top where a large bird daily ate away at his liver, which then grew back during the night. Asia might have been named for a queen. The Koran notes that Moses was raised by Asia, the wife of the Egyptian pharaoh, and that she was tortured because she believed in Moses' thoughts of God. Muhammad named Asia as one of four perfect women, and indicated that she ascended to heaven. Asia was also the name assigned to a nymph by Virgil. But, more likely, "Asia" derives from "asu," the Assyrian word for "east."

Africa, the third continent known to Europeans during ancient times, was so named because it was devoid, "a," meaning "without," of extreme cold, "frica," or frigidity. The Arctic, which is not one of the continents, took its name from the northern hemisphere's constellations,

Ursa Major and Ursa Minor, the Great and Little Bears. "Arkos" is the Greek word for "bear."

Antarctica was named because it was situated directly opposite the Arctic, and people at the time believed that its presence was necessary to stabilize the globe by counteracting the landmass of the Arctic. Australia, which is the smallest of the continents and the last to be named, having been previously called New Holland, comes from the Latin word "australis," meaning "south," because it resides solely south of the equator.

The two remaining continents, both in the Western Hemisphere, bear the name of one man. Thus, only one man's name is attached to any continent, and his name appears *twice*, in South America and North America. That one man, Amerigo Vespucci, from whose Christian name "America" was derived, is unrivaled in having his name perpetuated on the planet Earth.

Mankind has sought new horizons throughout history. The first European explorers probably set foot on North American soil over five hundred years before Vespucci was born. After the Norwegian Vikings had settled in Iceland and Greenland, Leif Erikson led a band of adventuresome men west across the Atlantic Ocean in the late tenth century. This occurred during a period of global warming. The land they reached, which is regarded to be L'Anse aux Meadows, Newfoundland, was assigned the name "Vinland" to denote the impressive number of wild grapes. The Vikings encountered Native American, whom they called *Skraelings*, meaning "wretches." After a period of time, during which the Vikings engaged in trade, they concluded that, although the land was bountiful, it was too dangerous to settle. They returned to their homes in Greenland and Iceland. The Little Ice Age that extended from 1000 to 1600 A.D. was perhaps a factor in the decline of the Vikings' exploration and absence of another European venture west to the area of original exploration for over five hundred years.

The saga of the naming of the Americas grew from the European discovery of the "New World" in the last decade of the fifteenth century. But the stimulus that led to that discovery was brought into focus two hundred years earlier. As the thirteenth century was drawing to a close, three men wearing strange garments debarked from a boat coming from Venice. Marco Polo, his father, and his uncle told an astonished audience that they were Venetians, who for over two decades had traveled over land and reached the court of Kubla Khan, the powerful ruler of China and most of the Far East. After serving in the court and noting the extraordinary riches of the kingdom, the Polos sailed past Zipangu (Japan), the Spice Islands, and through the

Persian Gulf before arriving home at Venice. Marco Polo's tales, which included the first mention in Europe of coal and paper money, in addition to fabulous wealth and a richness of spices, added impetus to the goal of those European powers with naval capability to reach eastward, to India, China, and the Spice Islands.

The quest was hampered and delayed by a series of impediments throughout the fourteenth century. The plague, the Hundred Years' War, and the religious schism consumed the attention of the European powers. Also, although the Venetian and Genoese naval presence dominated the Mediterranean Sea, formidable Islamic kingdoms dominated northern Africa and the Middle East, controlling trade in the region. Therefore, they required an alternative route.

Prince Henry of Portugal, commonly known as "The Navigator" because of his interest in navigation and his support of exploration, was the dominant ruler in Europe during the first half of the fifteenth century. He and several Portugese merchants sent ships south, reaching Senegal on the west coast of Africa by the mid-century mark. In 1486, Bartolomew Díaz sailed around the Cape of Good Hope at the southern tip of Africa, but had to return home before reaching his goal of India. In 1498, Vasco da Gama sailed all the way to India. But the journey around Africa was time-consuming and costly, and interest persisted in an alternative route, which was shorter, cheaper, and less dangerous.

On June 25, 1474, even before the accomplishments of Díaz and da Gama, Paolo dal Pozzo Toscanelli, a Florentine physician and humanist with an interest in astronomy and geography, wrote a letter directed to the King of Portugal. In this historic document, Toscanelli charted a course due west from Lisbon to the Chinese province of Mangi. The transoceanic distance was incorrectly estimated to be about 5,000 nautical miles, and the distance to Japan was a mere 2,000 miles. This error made the prospect of reaching the Far East by sailing westward even more attractive. But, in spite of this letter, Portugal persisted in the circum-African route to India, as evidenced by the subsequent Díaz and da Gama voyages.

By contrast, King Ferdinand and Queen Isabella of Spain, in an attempt to enrich, strengthen, and expand their realm, chose to focus on a westerly route to the Far East. They engaged the Genoese seaman, Christopher Columbus, to sail under the Spanish flag. Carrying Toscanelli's letter with him on his voyage, Columbus was the first to sail west across the Atlantic Ocean into the Western Hemisphere. On October 12, 1492, he made landfall on San Salvador or Watling Island, which the Indians called Guanahani. During this first voyage and the second voyage in 1493, Columbus landed on several other Carribean islands. On the third voyage, in 1498, Columbus and his crew landed on the South American continent at the Paria Peninsula of what is currently Venezuela, thus becoming the first European to set foot on that continent (see figure 1). It is possible that Columbus viewed the mouth of the Oronoco River. On his fourth and final voyage, Columbus explored the coast of Panama and Central America. One of the vessels that was scuttled during that voyage was discovered off the coast of Panama in 2001 and is currently the subject of an archeological project.

Although not well educated and certainly not a man of letters, Christopher Columbus was an extraordinary seaman, who, at times, sailed intuitively. His voyages afforded him a brief period of celebrity during which he was designated *Almirante del Mar Océano* and *Capitán General de la Armada* (Admiral of the Ocean Sea and Captain General of the Armada) by the Spanish sovereigns, Ferdinand and Isabella. The office of Admiral of the Ocean Sea gave Columbus jurisdiction over all Spanish ships sailing to or from the "Indies" west of the Azores.

Columbus persisted in the belief that he had reached Asia and the East Indies, accounting for his designation of the Natives he encountered as "Indians." But because he failed to discover precious metals or exotic spices, Columbus's glory turned into disgrace, and he died in Valladolid, Spain, in 1506, a neglected and almost forgotten man.

While Columbus was sailing in the Carribean under the Spanish flag, another Italian, Giovanni Gaboto, also known as John Cabot, was engaged in voyages of discovery for England. In 1497, one year before Columbus actually set foot on continental land in the Western Hemisphere, Cabot landed on Labrador and Newfoundland.

In spite of the historic background, in which it is generally accepted that Leif Erikson landed in the Western Hemisphere, and the unquestioned facts that Columbus was the first to land on the South American continent, and John Cabot the first to land on continental North America, not one of their names is appropriately recognized for their accomplishments.

Amerigo Vespucci (see figure 2), for whom North and South America are named, was a nebulous, ill-defined, and cryptic participant in the historic saga. He was born on March 9, 1452 (or 1454), the third child of ser Nastagio di Amerigo Vespucci, a minor notary in the Money-Changers Guild of Florence, Italy. Amerigo Vespucci, a contemporary of Niccolo Machiavelli, received his education, which included Latin, astronomy, and geography, mainly from his uncle, Georgio Antonio Vespucci, a distinguished scholar and monk in the order of San Marco. During his education, Vespucci met Paolo dal

Pozzo Toscanelli, the Italian physician who fostered Columbus's belief that a westward transoceanic voyage from Western Europe to China was both feasible and attractive because it encompassed a mere 5,000 nautical miles. During the period of his schooling, Amerigo Vespucci also developed a friendship with Piero di Tommaso Sonderini, who became Gonfaloniere of Justice of Florence, and to whom Vespucci would address his famous letters that played a role in the subsequent sequence of events.

Vespucci entered public life as a secretary serving Lorenzo di Pierfrancesco de' Medici, the cousin of Lorenzo the magnificent, who made Florence a center of Italian political and cultural power. After a brief diplomatic mission to France, Lorenzo di Pierfrancesco de' Medici sent Vespucci to Seville, Spain, where the Medici family had an interest in banks that helped finance early voyages of discovery and exploration. Vespucci entered the banking firm of Giannotto Berardi, who had formed a partnership with Christopher Columbus and participated in financing the first expedition. The company was expanded to include Vespucci himself, who first met Christopher Columbus after Columbus's triumphal return from his first voyage to the islands in the Caribbean Sea in 1492. The banking firm also helped finance Columbus's second voyage. Unfortunately, Berardi's bank was liquidated in 1495, and, although Vespucci was named executor of Berardi's estate, he became unemployed and destitute.

There is no reference to Amerigo Vespucci from 1495 until May 1499, when he joined a Spanish fleet commanded by Alonso de Hojeda, who had been dispatched by Cardinal Fonseca specifically to explore the Gulf of Paria and the coast of Venezuela, one year after Columbus's landing in the same region. Vespucci's role on this mission was inconsequential and ill defined. His participation earned him neither fame nor riches.

Vespucci's experience on that voyage, however, led the King of Portugal to enlist him as a pilot, astronomer, and cartographer on an expedition to the northern coast of Brazil, which was important in correcting Cabral's misconception that he had discovered an island, which he named "Terra Santa Cruz." The ships on the voyage in which Vespucci participated sailed along the east coast of South America almost as far as 50° South Latitude, and the exploration proved that the landmass was much larger than a simple island. Otherwise, this voyage, the chronicles of which rarely mention Vespucci, as was the case with his earlier expedition with de Hojeda, provided little improvement in Vespucci's fortunes.

On March 22, 1508, at age 57, Vespucci was appointed the first Pilot Major to the Casa de la Contratación (Board of Trade) in Seville.

In the Casa de la Contratación, the governing body of Spain's territories overseas, Vespucci, as the Pilot Major, was responsible for examination and preservation of sea charts, and instruction of pilots. In this capacity Vespucci drew up a royal *padron general,* a master chart of all the lands and islands of the Indies. The chart was routinely updated by additions and corrections brought back by explorers in order to provide a current inventory of lands discovered or claimed by Spain. But, despite Vespucci's position as Pilot Major, his estate remained meager, and, after his death on February 22, 1512, his widow found it necessary to petition the government for financial assistance.

The historical records pertaining to Vespucci's voyages to the Western Hemisphere are clear. The evidence of the truth rests mainly on copies of three letters Vespucci sent to his patron, Lorenzo de' Medici. The copy of the first letter was discovered by A. M. Bandini in 1745. F. Bartolozzi discovered the third letter's copy in 1789, and the copy of the second letter was uncovered by Count G. B. Baldelli Boni in 1827. These copies of the original letters were found in the collection of P. Vaglienti, who had collated reports dealing with the voyages of discovery.

The first letter, sent from Seville in July, 1500, concerns the voyage made between May 1499 and June 1500 by Alonso de Hojeda and Juan de la Cosa, who had sailed with Christopher Columbus and is credited as the author of the first map, a manuscript on oxhide dated 1500, to depict Columbus's discoveries and also present any part of the North American continent (see figure 10). This was Vespucci's first voyage, and his role was not defined. The letter is most pertinent in that it confirms that the expedition landed at the Paria Peninsula of Venezuela a year *after* Christopher Columbus had landed. The account corresponds to the two Spanish voyages, particularly the second voyage of the printed *Lettera.*

The other letters are not relevant to the primacy of the discovery of the South American continent. The second letter, dated June 4, 1501, and dispatched from Cape Verde, refers to the first part of the Portugese voyage that explored the northern coast of Brazil. This voyage was the one detailed in the 1503 printed published document entitled *Mundus Novus* (The New World). The third letter is undated, but was certainly sent from Lisbon after Vespucci returned to Portugal from his "first Portugese" voyage. The letter describes twenty-seven days spent with a Brazilian tribe and suggests that the expedition reached 50° South Latitude, which would not be reached (and surpassed) again until the Magellan expedition in 1520.

Unlike Columbus, Amerigo Vespucci never led an expedition of discovery and never achieved fame during his life. It is therefore truly extraordinary and enigmatic that a man who personally claimed no primacy should have his name not only become part of, but dominate, an international geographic lexicon. The sequence and circumstances that resulted in this unique recognition of a man's given name are most readily analyzed by fitting together the pieces of an historical jigsaw puzzle. Individual pieces offer suggestions but no solution. The whole picture, however, when completed by interlocking the individual pieces, does offer a credible interpretation.

Vespucci's three manuscript letters establish the fact that the de Hojeda expedition landed on the Paria Peninsula a year after Columbus. They also establish that Vespucci never set foot on the North American continent. These letters lay hidden in storage for almost three hundred years, and were not available to scholars in the sixteenth century. In their place, two other textual documents were published and set the stage for eventually imprinting Vespucci's name on the two continents in the Western Hemisphere.

The introduction of letter printing from movable type by Johannes Gutenberg in Mainz, Germany, about 1454 or 1455, led to a broader and more rapid dissemination of information. Initially, the printing press brought forth textual material; subsequently the process was applied to graphics, including maps. The earliest extant printed map is a crude woodcut mediaeval T-O diagram that appeared in an edition of Isidore of Seville's famous medieval encyclopedia, printed in Augsburg in 1472. The first printed atlas was an edition of Ptolemy's *Cosmographia*, published in Bologna in 1477.

In the case of the designation of Vespucci as the discoverer of the New World, the same sequence of textual material followed by graphic representation pertained. The first printed piece, a short letter of only four pages, bearing the title *Mundus Novus*, the "New World," was published in Florence, Italy in 1503, only a half century after Gutenberg's invention of movable type and a printing press. The letter was allegedly a Latin translation made by the Veronese humanist, Father Giovanni del Giacondo, from a lost and never found Italian letter. Supposedly, the original letter, which bore the name "Albericus Vesputius" as correspondent, had been sent from Lisbon by Vespucci to his patron, Lorenzo de' Medici. The brief document detailed a voyage, sponsored by King Manuel I of Portugal, that departed from Lisbon on May 14, 1501, and returned more than a year later, on July 22, 1502. The letter noted that the expedition landed somewhere between Venezuela and the northern coast of Brazil on August 7, 1501, and continued along the coast to 50° South Latitude.

Emulating Columbus's famous letter, *De insulis in mari Indico nuper inventis* of April 29, 1493, in which Columbus announced his discoveries, the "Vesputius" letter described geographic and ethnographic features of the area explored, including the social and sexual activities of the Natives. The "Vesputius" letter was distinctive and particularly captivating because, unlike the Columbus letter, it referred to the newly discovered lands, for the first time, as *Mundus Novus*—a New World!

A translation of the first paragraph states:

> In the past I have written you in rather ample detail about my return from those new regions which we searched for and discovered with the fleet, at the expense and orders of His Most Serene Highness the King of Portugal (Manuel I), and which can be called a New World, since our ancestors had no knowledge of them and they are entirely new matter to those who hear about them. Indeed, it surpasses the opinion of our ancient authorities, since most of them assert that there is no continent south of the equator, but merely that sea which they call the Atlantic; furthermore, if any of them did affirm that a continent was there, they gave many arguments to deny that it was habitable land. But this last voyage of mine has demonstrated that this opinion of theirs is false and contradicts all truth, since I have discovered a continent in those southern regions that is inhabited by more numerous peoples and animals than in our Europe, or Asia, or Africa, and in addition, I found a more temperate and pleasant climate than in any other region known to us, as you will learn from what follows, where we shall briefly write only of the main points of the matter, and of those things more worthy of note and record, which I either saw or heard in this new world, as will be evident below.

This monumental declaration of a New World, combined with the description of an idyllic and bountiful land with enticing Natives—in all, a Paradise on earth—received immediate and broad attention. The letter underwent several editions and was quickly reprinted in Venice, Paris, Antwerp, and several cities in Germany.

The excitement generated by the declaration of the discovery of a New World was amplified by the only other contemporary printed account of Vespucci's expedition to the Western Hemisphere. This second document was printed first in Florence in 1505 and was entitled: *Lettera di Amerigo Vespucci delle isole nuovamente trovate in quattro suoi viaggi* (Letter of Amerigo Vespucci concerning the isles newly discovered on his four voyages). This 32-page letter was originally written in 1504, as indicated by the notation at the end of the printed work:

"Data in Lisbona a di 4, septembre 1504. Servitore Amerigo Vespucci in Lisbona." The letter was addressed to a "Magnificent Lord," who has been identified as Vespucci's childhood friend, Piero di Tommaso Soderini, Gonfaloniere of Justice, the head of the Florentine government. Doubtless Soderini became the addressee because Vespucci's patron, Lorenzo de' Medici, had died the year before.

The thirty-two pages of the *Lettera*, added to the previously published four pages of *Mundus Novus,* magnified Vespucci's reputation exponentially. But it actually detailed a fraud! The *Lettera* described four specific journeys to the Western Hemisphere, the first two under the flag of Spain and the last two under the flag of Portugal. The *Lettera* claimed that the first of these voyages took place between May 10, 1497, and October 15, 1498, and included explorations of the Paria Peninsula of Venezuela. This claim clearly challenged Columbus's primacy of discovery by stating that Vespucci landed on the South American continent a year *earlier* than Columbus (at approximately the same time that John Cabot landed in Labrador and Newfoundland). This voyage eventually became the center of controversy. It totally misrepresented the fact that the voyage to the Paria Peninsula of Venezuela took place in 1499, one year *after* Columbus's landing at that region, and that Vespucci played an insignificant role under the leadership of Alonso de Hojeda on the voyage.

The second voyage detailed in the letter reportedly took place between May 16, 1499, and September 8, 1500, and extended along the coast of South America from Venezuela to Cape San Augustin (8° South Latitude) at the easternmost tip of Brazil.

This report actually corresponds with Vespucci's first trip with de Hojeda, although there is a question as to whether that trip extended quite so far south.

The third voyage in the *Lettera* constituted the first of the Portugese voyages referred to in the manuscript letters and took place between May 14, 1501, and July 22, 1502. It corresponds to the voyage detailed in the *Mundus Novus* of "Albericus Vesputius," but also includes a description of the discovery and naming of Cape San Augustin and Bahai de Todos os Santos.

The fourth and final voyage described in the *Lettera* never took place at all. The *Lettera* indicates that this voyage extended from May 10, 1503, to July 18, 1504, but Vespucci could not have sailed to the Western Hemisphere after 1502, when it is known that he returned to Spain permanently.

Thus, the printed *Lettera* clearly distorted all of Vespucci's actual voyages to the Western Hemisphere, ascribing an erroneous date to the first, exaggerating the others, and fabricating the fourth (see figure 3)!

Nevertheless, with Vespucci's three manuscript letters in an obscure archive hidden from the public, the two printed documents, *Mundus Novus* and *Lettera*, formed the textual basis for subsequent recognition of Amerigo Vespucci as an explorer of great importance.

Two publications in 1507 increased Vespucci's reputation, a 126-page anthology of travels and a new description of the world's geography. The anthology, entitled *Paesi nuovamente retrovati et Novo Mondo da Alberico Vesputio Florentino intitulato* (Newly discovered countries and New World of Alberico Vesputio Florentine), was edited by Fracanzano da Montalboddo, a professor of rhetoric at Vicenza, where it was printed on November 7, 1507. The book, which was reprinted several times and translated into Latin, described several Portugese expeditions, the first three voyages of Columbus, and the *Mundus Novus* of Vespucci. The title's inclusion of Vespucci's name and exclusion of Columbus's name confounded its statement, which could be interpreted as indicating that Vespucci discovered the new countries. It was the geography, which was published a few months earlier, however, that had the definitive effect, and was the decisive factor in declaring Vespucci's primacy of discovery of the "Americas."

In the small town of Saint-Dié, in the Vosges Mountains, in the no longer extant Duchy of Lorraine, the Gymnasium Vosagense (Gymnase Vosgien) was established to spread knowledge, primarily by printing books. Saint-Dié had previously played a small role in the voyages of discovery. In 1410, at Saint-Dié, Cardinal Pierre d'Ailly composed *Ymago Mundi* (Image of the World), which was published in Louvain, Belgium in 1483. The annotated copy that Columbus carried with him during his voyages has been preserved in the Archivio General de Indias in Seville.

Early in the sixteenth century, a little more than half a century after Gutenberg's famous Bible came off the press at Mainz, Gaultier Lud, a printer and secretary to René II, the reigning Duke of Lorraine, who also ceremonially bore the grandiose but meaningless titles of "King of Jerusalem and Sicily" and "Count of Provence," organized the Gymnasium Vosagense. Canon Lud gathered around him monks and laymen, learned humanists, editors, translators, and illustrators, with the specific goal of producing meaningful books. The literary circle at the Gymnasium included Nicholas Lud (Gaultier's nephew), Johannes Basinus, Mathias Ringmann ("Philesius"), and Martin Waldseemüller ("Hylacomylus"), the most influential figure in the naming of America.

Waldseemüller is thought to have been born at Radolfzell on the shore of Lake Constance between 1470 and 1475. In 1490, he matriculated at the University of Freiburg in Germany, where he studied theology. At the conclusion of his studies, Waldseemüller became a clergyman in the diocese of Constance, and later he occupied the position of Canon at Saint-Dié until his death on March 16, 1519. In addition to his religious devotion, Waldseemüller had a great interest in geography and cartography. In 1505, he spent time in Strasbourg studying geography, especially the classic maps of Ptolemy, a Greek scientist and geographer who lived in Alexandria, Egypt, during the second century A.D. Ptolemy's maps were rediscovered during the twelfth century and translated into Latin in 1406.

Although the first printed editions of Ptolemy's geography were not published until 1475 (without the maps) and 1477 (with maps), his maps had a significant influence of the European view of the world. The important first German edition with five new maps was published in Ulm in 1482. In Strasbourg, in 1513, an edition of Ptolemy's geography that included twenty new maps, among them the first maps of the New World to be included in an atlas, was published. The maps are credited to Waldseemüller.

On April 25, 1507, six year's before Waldseemüller's edition of Ptolemy appeared, the Gymnasium produced a small book that effectively and permanently inscribed the name "America" on the cartography and geography of the world. The fifty-two-page book was entitled *Cosmographiae introductio, cum quibusdam geometriae ac astronomiae principiis ad eam rem necessariis Insuper quatuor Americi Vespucij navigationes. Universalis Cosmographiae descriptio tam in solido quam plano. eis etiam insertis, que Ptholomaeo ignota a nuperis reperta sunt.* (Introduction to cosmography, with the necessary fundamental principles of geometry and astronomy. Included are the four voyages of Amerigo Vespucci. And also a description of the universe in a spherical and flat form of parts unknown to Ptolemy and recently discovered.)

There are two known editions of *Cosmographiae Introductio,* which were printed at the Gymnasium Vosagense in Saint-Dié. In the first edition, Martinus Ilacomilus (a graecized form of Martin Waldseemüller, meaning "miller of the forest lake,") was named as the editor, while the second edition, printed on August 29, 1507, credits the Gymnasium Vosagense as the editor. Certainly, Waldseemüller deserves full credit, because he not only wrote the cosmography but also drew the two important maps.

As the title page specifically indicates, the book consists of four parts: a geographical introduction, an account of Vespucci's alleged four voyages (as outlined in the *Lettera*), a gore map that is drawn and cut in such a way that it can be affixed to the surface of a globe, and a flat map. Not only is Vespucci's name included in the subtitle, but the text strongly embellishes his status. The text declares that Vespucci expanded the Ptolemeic world and recognizes him as the discoverer of new lands: "et maxima pars Terrae semper incognita nuper ab Americo Vesputio respartae" (and the larger part of the land always unknown until discovered by Amerigo Vespucci).

The most forceful declaration of Vespucci's accomplishments appears in chapter 9 of the *Cosmographiae Introductio*, where, after describing the geography of Europe, Africa, and Asia, Waldseemüller writes:

> Now these parts of the earth have been more extensively explored and a fourth part has been discovered by Amerigo Vespucci (as will be set forth in what follows). Inasmuch as both Europe and Asia received their names from women, I see no reason why any one should justly object to calling this part Amerige, i.e., the land of Amerigo, or America, after Amerigo, its discoverer, a man of great ability. Its position and the customs of its inhabitants may be clearly understood from the four voyages of Amerigo, which are subjoined.

And to emphasize further this verbal assertion, in the margin of the page on which the text is printed, the word "America" appears. Nowhere in the printed document does Columbus's name appear!

Returning to the assertion that "a picture is worth a thousand words," it was Waldseemüller's *en plano* (flat map), which appeared in the *Cosmographiae,* that placed Vespucci's name on land in the "New World." The map was lost to the public for almost four centuries, until 1900, when Professor Joseph Fischer, a Jesuit priest, discovered a long-sought-for copy in the Castle Wolfegg in Württemberg, Germany. This only extant copy of the map is a woodcut, consisting of twelve sections arranged in three zones; each of the four sections measures 23.5 x 17.5 inches (see figure 4). The overall dimensions of the map are 54 x 73.5 inches, about 36 square feet. Inscribed on the lower edge is its title: UNIVERSALIS COSMOGRAPHIA SECVNDVM PTHOLOMAEI TRADITIONEM ET AMERICI VESPVCCI ALIOR_QVE LVSTRATIONES.

The map depicts the earth in a modified Ptolemaic coniform projection with curved meridians. Two small maps depicting the Eastern and Western Hemispheres are positioned above the main map. Each of these small maps is flanked by a portrait of the man associated with that hemisphere. To the left is the Eastern Hemisphere or the Ptolemaic world and a portrait of "Claudii Phtolomei." To the right is a map of the Western Hemisphere or New World and a portrait of

"Americi Vespucci" (see figure 4a). What is more, the words "Americi Vespucci" have been printed larger than "Claudii Phtolomei." On the large main map, in the southern half of the Western Hemisphere, in the middle of a continental landmass, the word "America" appears (see figure 5).

The second map referred to on the title page of *Cosmographiae Introductio* is the "solido" or gore map to be placed on a globe (see figure 6). In the text of the book, it is stated: "We have therefore arranged matters so that in the plane projection we have followed Ptolemy as regards the new lands and some other things, while on the globe, which accompanies the plane, we have followed the description of Amerigo that we subjoin."

In 1950, at the Parke-Bernet Galleries in New York City, an exciting sale took place. One of only two known copies of this first globular map of the world to include the Western Hemisphere, on which the word "America" appears, was purchased by the James Ford Bell Collection of the University of Minnesota from the Hauslab-Leichtenstein collection. The map measures a mere 9.5 x 15 inches, and was cut from a single wood block for printing. In the Western Hemisphere, the globe gores depict parallels that correspond to contemporary Portugese and Spanish sea charts, and, like the flat map, the word "America" is inscribed in the southern continent in the Western Hemisphere. No other words appear on that continent.

And so it was that, in 1507, both in his text and on two very different maps, Martin Waldseemüller named the southern continent of the Western Hemisphere "America"!

It is probable that this misnaming of America can be attributed to the errors perpetrated by the two widely distributed books, *Mundus Novus* and *Lettera.* But, shortly after publication of the *Cosmographiae*, Waldseemüller apparently changed his mind about the naming of America. He came to realize that Christopher Columbus, and not Amerigo Vespucci, was the true discoverer of the South American continent. Perhaps it became apparent to him that the works that had played a role in his original conclusion were inaccurate or complete forgeries.

Waldseemüller apparently recanted six years later, when he produced his own edition of the Ptolemy Atlas, published in Strasbourg in 1513. On the first map devoted to America to appear in any atlas, the word "America" does not appear. The map entitled "Tabvla Terre Nove" (Chart of the New Land) has the South American continent designated "Terra Incognita" (Unknown land) (see figure 7). That land bears the statement: "The lands and adjacent islands were discovered by Columbus sent by authority of the King of Castile." In this atlas, there is no mention of Amerigo Vespucci, nor is there any reference to "America" on the map of the New World, on an included world map, or anywhere in the text.

Similarly, the word "America" is absent from the world map published in 1515, in Strasbourg, by Gregor Reisch in his *Margarita philosophica nova.* On that map, the words "PARIA SEVPRISILIA" denote the part of the South American continent that is shown. The same year, however, on the so-called "Paris" or "Green" Globe, the name "America" appears in four places, and it is placed for the first time on the North American continent. In 1516, Waldsemüller published another world map, "Carta Marina Navigatoria Portvgallen Navigationes Atque Tocius Cogniti Orbis Terre Marisque . . . ," which was the first printed nautical chart of the modern world; it was found in the same binding with his 1507 maps. This twelve-sheet tour de force, which measures 8 x 4.5 feet, accords both Christopher Columbus and Amerigo Vespucci recognition, but the main names on the South American continent are "TERRA NOVA" and "TERRA PAPAGALLI," meaning "land of parrots."

In 1520, a map published by Peter Apian in Vienna boldly imprinted the word "America" across the southern continent of the Western Hemisphere. Apian's map was a frank plagiarism of Waldseemüller's flat map from *Cosmographiae Introductio.* Over the ensuing several years, on maps by Laurent Fries (Strasbourg 1522), Franciscus Monachus (Antwerp ca. 1527) Peter Apian (Ingolstadt 1530), Johann Honter (Cracow 1530), Oronco Finé (Paris 1531), and Sebastian Münster–Hans Holbein? (Basle 1532), the word "America" appears on the South American continent. Thus, despite the fact the Waldseemüller himself realized his mistake and tried to correct it, the idea that the New World should bear the name "America" persisted and grew. The fate of the name was sealed in 1538, when Gerard Mercator, the renowned cartographer, who later developed a more accurate method for representing the globe's curvature on a flat surface—the Mercator projection—added his imprimatur by inscribing the word "America" on both the northern and southern continent in the Western Hemisphere (see figure 8).

Although his name had been attached to a continent in the New World in 1507, Amerigo Vespucci was a relatively inconsequential bureaucrat in the employ of the King of Spain, without wealth, without glory or fame, when he died five years later in Seville. Christopher Columbus, whose name should have been placed on a map based on primacy, rested almost completely forgotten by his contemporaries in a grave in Valladolid, Spain, having died in 1506, the year before Waldseemüller published his most influential maps.

During their lives, the two protagonists in the drama of vying for recognition based on primacy of discovery and the right to have his name appended to two continents in the New World actually had little to do with the controversy and maintained an amicable and mutually respectful relationship as evidenced by Columbus's correspondence.

On February 5, 1505, one year before he died, Christopher Columbus wrote his son a letter, which is preserved in the Columbus Library.

My Dear Son,

> Diégo Mendez left here on Monday, the 3rd of this month. Since his departure, I have talked with Amerigo Vespuchy, who is on his way to the Court, where he has been called to be consulted on several points connected with navigation. He has at all times shown the desire to be pleasant to me and he is a very respectable man. Fortune has been adverse to him, as to many others. His efforts have not brought him the reward he might by rights have expected. He leaves me with the desire to do me service, if it should be in his power. I am unable here to point out in what way he could be useful to me, because I do not know what might be required at Court; but he is decided to do all in his power on my behalf. You will see in what way he can be employed. Think the matter over, as he will do everything, and speak, and put things in train; but let all be done secretly, so as not to arouse suspicion of him. I have told him all I can about my affairs, and of the payments that have been made to me and are due. This letter is also for Adelantado, for he can see in what way use can be made of it, and I will apprise you of it, etc., etc.

Dated in Seville the 5th of February (1505)

Clearly, Columbus held no ill will toward Vespucci, and he, personally, did not regard *Mundus Novus* or the *Lettera,* which were published during Columbus's lifetime, as a false claim by Vespucci to aggrandize fame. Neither man held any animosity toward the other. But, about fifty years after Vespucci's death, a bitter conflict erupted between the two explorers, now dead, in a manner not unlike a puppet show in which invisible puppeteers jerk the marionettes through their movements.

The first input came at a time when the English, French, and Dutch were evidencing designs for the Western Hemisphere, and

Spain had the need to reassert Columbus's (and therefore Spain's) primacy and claim to all of the continental lands in that hemisphere. The initial declarations came from the pen of Bishop Bartolomé de Las Casas, a distinguished clergyman and historian who was well acquainted with the new world. Las Casa was born in 1474 and went to Hispanola in 1502 where, having been the first man to be ordained as priest in the New World, he served as a priest and eventually became a bishop. Known as the "Apostle of the Indians," Las Casas was passionately devoted to the plight of the oppressed Natives, and, in 1517, he sailed to Spain to plead their case to Carlos I. Beginning in about 1527 and occupying most of his life, Las Casas worked on his lengthy tome, *Historia de las Indias* (History of the Indies).

In the 1550s, he returned to Spain permanently, and at age 85 he continued writing his epic until his death in 1566. Fittingly, the scholarly writing took place at the College of San Gregorio in Valladolid, the city in which Columbus died. Las Casas had available for his use all of Columbus's papers, including copies of his travel journals. The manuscript for *Historia de las Indias*, which includes an abstract of Columbus's first voyage and is preserved in the National Library in Madrid, certainly was influential in its time, but the text was not printed until 1875. The great American historian, Samuel Eliot Morison, stated that the one book on the discovery of America that he should wish to preserve if all others were destroyed was *Historia de las Indias.*

Las Casas, whose father, Pedro, participated in Columbus's second voyage, ignited the cause directed against Vespucci, who Las Casas claimed was guilty of stealing glory from Columbus, for whom the continent should have been justly named. The Bishop did have relations with the Colon family, Columbus's descendants, and the text offered support for the legal action Columbus's heirs were taking against the Treasury (Fiscal) regarding the observance of Columbus's privileges. In defense of the vitriol that issued forth from Las Casas's pen, it should be noted that he had only a Latin version of the *Lettera,* in which Vespucci allegedly chronicled four voyages, as his source, and Las Casas was unaware that it was Waldseemüller, not Vespucci, who was responsible for imprinting the name on the continent.

Las Casas's own words in chapter 140 best summarize his position:

> It is manifest that the Admiral Don Cristóbal Colón was the first by whom Divine Providence ordained that this our great continent should be discovered, and chose him for the instrument through whom all these hitherto unknown Indies should be shown to the world. He saw it on Wednesday, the 1st of August, one day after he discovered

> the island of Trinidad, in the year of our salvation 1498. He gave it the name Isla Santa, believing it was an island. He then began to enter the Gulf of La Bellena, by the entrance he called the mouth of the Serpent, finding all the water fresh, and it is this entrance that forms the island of Trinidad, separating it from the mainland called Santa. On the following Friday, being the 3rd of August, he discovered the point of Paria, which he also believed to be an island, giving it the name of Gracia. But all was mainland, as in due time appeared, and still more clearly now is it known that here is an immense continent.
>
> It is well here to consider the injury and injustice which Americo Vespucci appears to have done to the Admiral, or that those have done who published *Four Navigations*, in attributing the discovery of the continent to himself, without mentioning anyone but himself. Owing to this, all the foreigners who write of those Indies in Latin, or in their own mother tongue, or who make charts or maps, call the continent America, as having been first discovered by Americo.
>
> For as Americo was a Latinist, and eloquent, he knew how to make use of the first voyage he undertook, and to give the credit to himself, as if had been the principal captain of it. He was only one of those who were with the captain, Alonso de Hojeda, either as a mariner, or because, as a trader, he had contributed the expenses of the expedition; but he secured notoriety by dedicating his *Navigations* to King René of Naples. Certainly these *Navigations* unjustly usurp from the Admiral the honor and privilege of having been the first who, by his labors, industry, and the sweat of his brow, gave to Spain and the rest of the world a knowledge of this continent, as well as of all the Western Indies. Divine Providence reserved this honor and privilege for the Admiral Cristóbal Colón, and for no other. For this reason no one can presume to usurp the credit, nor to give it to himself or to another, without wrong, injustice, and injury committed against the Admiral, and consequently without offence against God.

Las Casas goes on to prove that the voyage commanded by Alonso de Hojeda took place a year after Columbus had landed on the same region of the Paria Peninsula. Las Casas submitted as evidence de Hojeda's testimony, which had been given in an unusual trial brought by the heirs of Columbus against the Treasury of Spain in 1516. Not one of over a hundred witnesses at the trial disputed de

Hojeda's claim that Columbus had been the first European to set foot on the South American continent. In addition, Las Casas pointed out that in the original Italian edition of the *Lettera*, describing Vespucci's alleged first voyage of 1497—the one on which primacy was based—Vespucci is said to have landed at a place called "Lariab," not Paria. It was later shown that a printing change was made in the Latin edition, converting the word to "Paria."

These errors in dating, in translation, and in simple interpretation—some of which might have been perpetrated by Vespucci (though probably not)—were widely disseminated and perpetuated by works in print. The true power of the printed word and the printed image can be seen by the fact that the testimony of de Hojeda, taken in 1516, was insufficient to affect popular consciousness. In spite of the evidence, maps and other publications continued naming the New World the Americas.

Although Las Casas, both in his own time and later, had little impact on the naming of America, the desire to reclaim Columbus's primacy of discovery continued. In 1601, Antonio de Herrera y Tordesillas, Historiographer of the Indies under Phillip II of Spain, published his *Descriptión de las Indias Occidentales.* In it, he emphasized the primacy of Columbus's discovery of the New World. Building on La Casas's work, which at the time was still unpublished but available in manuscript, Herrera reiterated the point that the dates of Vespucci's voyages as outlined in the *Lettera* were incorrect and that Vespucci had not sailed to the New World until 1499, and then under the command of de Hojeda. Herrera specifically suggested that Vespucci had intentionally falsified the dates of his voyage in an effort to glorify himself at the expense of Columbus. In 1627, Fray Pedro Simon went so far as to declare emphatically that all works and all maps using the name "America" should be suppressed.

Outrage over the naming of America continued to grow. More than two centuries later, in 1856, one of America's most distinguished thinkers and essayists, Ralph Waldo Emerson, indicated in his book *English Traits* that the naming of America after Amerigo Vespucci was as ridiculous as England's adoption of Saint George as its patron saint.

> George of Cappadocia, born at Epinania in Cilicia, was a low parasite who got a lucrative contract to supply the army with bacon. A rogue and informer, he got rich and was forced to run justice. He saved the money, embraced Arianism, collected a library, and got promoted by a faction to the episcopal throne of Alexandria. When Julian came, A.D. 361, George was dragged to prison; the prison was

> burst open by the mob and George was lynched, as he deserved. And this precious knave became, in good time, Saint George of England, patron of chivalry, emblem of victory and civility, and the pride of the best blood of the modern world.
>
> Strange that the solid truth-speaking Briton should derive from an imposter. Strange that the New World should have no better luck, that broad America must wear the name of a thief. Amerigo Vespucci, the pickle-dealer at Seville, who went out, in 1499, a subaltern with Hojeda, and whose highest naval rank was boatswain's mate in an expedition that never sailed, managed in this lying world to supplant Columbus and baptize half the earth with his own dishonest name. Thus nobody can throw stones. We are equally bad off as our founders; and the false pickle-dealer is an offset to the false bacon-seller.

With all due respect to Emerson's brilliant mind and poignant prose, Vespucci was thrice maligned by a man whose poems, orations, and essays are landmarks of American thought and literary expression. First, Vespucci was not a thief; he personally played no role in detracting from Columbus's glory and accomplishments. Indeed, according to Columbus's own letter, it would seem that Vespucci was a supporter of Columbus. Second, a respected banker and financier of voyages of exploration hardly merits classification as a "pickle-dealer." Third, Vespucci was in no way involved in the baptizing of half the earth.

The naming of America was stimulated by whoever wrote *Mundus Novus* and the *Lettera*; enforced by the vagueness and misinterpretation of the anthology, *Newly Discovered Countries and New World of Alberico Vesputio, Florentine*; and was finalized and ingrained by the text of *Cosmographiae Introductio* and Waldseemüller's two 1507 maps. Vespucci almost certainly was not the author of *Mundus Novus* or the *Lettera* and could not have contributed to Waldseemüller's textual and cartographic works.

Vespucci himself remained innocent and probably ignorant of the dissemination of the various narratives and geographies that attributed the discovery to him. There is no evidence that he ever claimed primacy of discovery. Vespucci's own manuscript letters present the facts honestly, and include dates that correspond to the true dates of his voyage with de Hojeda, which do not infringe upon the fact that Columbus was the first to set foot on the South American continent.

What may be attributed to Vespucci is the conviction that the land he reached was a New World. Christopher Columbus, a talented seaman and intrepid explorer, but poorly educated, was certainly the first European to cross the Atlantic Ocean and set foot on continental land in the Western Hemisphere. However, he persisted in the belief that he had reached Asia and the Indies. Vespucci, who had no distinction as a seaman, was well educated and a financier. He held the conviction that the newly discovered land to the east was distinct from Asia and the Indies, a conviction made possible by southern expeditions that revealed the vast extent of the landmass they had encountered.

Since the naming of the northern and southern continents in the Western Hemisphere "America" is based on errors in dating and interpretation, and, perhaps, on editorial license, the beguiling question, then, is, in a perfect world with perfect knowledge, "What should these continents have been named?" And "What would be a more appropriate name for our country?"

Many names that derive from the Native Americans were adopted by the explorers. Some remain, designating mountain ranges, such as Appalachian, and lakes, such as Ontario, Erie, Huron, Michigan, and Okechobee. Many native words designate states, such as Alabama, Arizona, Arkansas, Connecticut, the Dakotas, Idaho, Illinois, Indiana, Iowa, Kansas, Kentucky, Massachusetts, Minnesota, Mississippi, Missouri, Nebraska, Ohio, Oklahoma, Tennessee, Utah, Wisconsin, and Wyoming; and countless cities adopted names derived from many of the Native American languages. But the naming of a continent that represented the possibility for expansion of European interest and authority was seen by each of the European powers as a cartographic expression of power.

Based on primacy of European discovery, the southern continent obviously should be named after Christopher Columbus. If past traditions were to be honored, as Waldseemüller suggested in his *Cosmographiae Introductio,* because two previously named continents in the Old World bear the given names of women, then Columbus's Christian name, Christopher or Cristoforo, would be preferable. But other choices derived from his family name would also be acceptable and perhaps a little less awkward: Columbus, Columbo, Columbia, or Colon.

Alas, all that remains of the name of the "Admiral of the Ocean Sea" on a map is Colombia: the name of a modest-sized South American country, the name that was assigned to the capital of South Carolina in 1786, and the name given in 1792 to the district in which the capital of the United States of America is located. Both of the latter names derive from a poem written at the beginning of the American Revolution by Philip Freneau, "American Liberty," which credited Columbus as the progenitor of the emerging nation. The first important use of the word "Columbus" itself was by the newly designated capital of the state of Ohio in 1812.

If the southern continent in the Western Hemisphere should have been named for Columbus, for whom should North America have been named? Neither Columbus nor Vespucci ever had anything to do with this continent. Neither of them discovered it; neither set foot on it; neither even knew it existed. But since the southern and northern continents are connected by an isthmus and Columbus merits primacy for discovering continental land in the western hemisphere, North and South Columbo or Columbia would not have been unreasonable.

But, if naming is based on primacy, then perhaps Leif Erikson should be acknowledged and the northern continent should bear the name that he assigned, Vinland. Arguing against this is the fact that the Vikings did not regard their discovery to be a new continent but rather just another island similar to Iceland or Greenland. Also, they were unable to establish any long-term settlements on the land.

Since the naming emanated from Europe, Giovanni Gaboto or John Cabot, who was the first Renaissance European to set foot on land in the northern part of the Western Hemisphere, could have provided an historically appropriate name—Cabotia or Gaboto. Primacy of discovery combined with the fact that an isthmus connects the two continents would result in North and South Cabotia.

Focusing on the United States, because Ponce de Leon was the first European to touch our soil, when he landed on and named Florida in 1513, our country could honor him. But his name is absent from our continental land, and is memorialized only by the name of the second largest city in Puerto Rico, the island which he governed. In 1521, Francisco Gordillo and Pedro de Quexos made a landfall at what is now Winyah Bay, South Carolina, but there were no consequences of their exploration, and their names appear on no maps. Giovanni da Verrazzano was the first European to define the east coast of our country during his voyage of 1524, extending from northern Florida to Labrador. His name is absent from the geography of the United States, and has been relegated to a suspension bridge crossing the entrance to New York Bay, which he was the first to explore.

Other errors have been made regarding the name "America." If Vespucci's Christian name was deemed to be appropriate then we would have South Amerigo. If the Latin equivalent was considered to be more in keeping with the enormity of the discovery, Albericus would have been the correct translation, not Americus. Certainly, South Alberica is harsher on the ear than South America.

"Vespucci" as a place name remains absent, but Vespucci continues to be inappropriately maligned for a personal role in a deliberate aggrandizement of fame. Ogden Nash, in his distinctive style, addressed the misnaming of America in the poem "Columbus":

> So Columbus said, Somebody show me the sunset and somebody did and he set sail for it,
> And he discovered America and they put him in jail for it,
> And the fetters gave him welts,
> And they named America after somebody else,
> So the sad fate of Columbus ought to be pointed out to every child and every voter,
> Because it has a very important moral, which is, Don't be a discoverer, be a promoter.

The "United States of Cabotia," "Vinland the Beautiful," "The Columbian Way" all would have been alternate expressions. But America by any other name would still be America!

References

Ailly, Pierre d', Cardinal. *Ymago Mundi.* Louvain, 1483.

Baldelli Boni, Count G. B. *Il Milione of Marco Polo.* Florence: Da'torchi de G. Pagani, 1827.

Bandini, A. M. *Vita et Lettere di Amerigo Vespucci.* Florence, 1745.

Bartolozzi, F. *Richerche istorico-critiche circa alle scoperte di Amerigo Vespucci.* Florence, 1789.

Columbus, Cristoforo. Letter, April 29, 1493: *De insulis in mari Indico nuper inventis.*

Emerson, Ralph Waldo. *English Traits.* Boston: Phillips, Samson, and Co., 1856.

Fischer, Joseph, and Von Weiser, Franz. *Die älteste Karte mit dem Namen Amerika aus dem Jahre 1507 und die Carta Marina aus dem Jahren 1516 des M. Waldseemüller (Ilacomilus).* Innsbruck: Wagner, 1903.

———. *The Cosmographiae Introductio of Martin Waldseemüller in Facsimile.* Edited by Charles George Herbermann. New York: United States Catholic Historical Society, 1907.

Harrisse, Henry. *The Discovery of North America.* London, 1892. Reprint, Amsterdam: N. Israel, 1961.

Herrera y Tordesillas, Antonio de. *Descriptión de las Indias Occidentales.* Madrid, 1601.

Isidore of Seville. *Isidorus Hispalensis Etimologiarum.* Augsburg, 1472.

Las Casas, Bartolomé. *Historia de las Indias.* Madrid: Impr. de M. Ginesta, 1875.

Magnaghi, A. "Una supposa lettera inedita di Amerigo Vespucci sopra il suo terzo viaggio." *Bolletino della R. Societa Geografica Italiana*, Ser. 7, vol. 2 (1937): 589–632.

Nash, Ogden. "Columbus." In *The Face is Familiar.* Boston: Little, Brown and Company, 1940.

Ptolemy. *Cosmographia.* Vicenza, 1475.

———. *Cosmographia.* Bologna, 1477.

———. *Cosmographia.* Ulm, 1482.

———. *Geographia.* Edited by Martin Waldseemüller. Strasbourg, 1513.

Vespucci, Amerigo. *Lettera di Amerigo Vespucci delle isole nuovamente trovate in quattro suoi viaggi.* Florence, 1505.

———. *Mundus Novus.* Florence, 1503.

———. *Paesi nuovamente retrovati et Novo Mondo.* Edited by Fracanzano da Montalboddo. Vicenza, 1507.

Waldseemüller, Martin. *Cosmographiae introductio, cum quibusdam geometriae ac astronomiae principiis ad eam rem necessariis Insuper quatuor Americi Vespucij navigationes. Universalis cosmographiae descriptio tam in solido quam plano eis etiam insertis, quae Ptholomaeo ignota a nuperis reperta sunt.* Saint-Dié, 1507.

SELECTED READINGS

Lester, C. Edwards. *The Life and Voyages of Americus Vespucci: With Illustrations concerning the Navigator, and the Discovery of the New World.* Fourth edition. New Haven, Conn.: Horace Mansfield, 1853.

Letters from a New World: Amerigo Vespucci's Discovery of America Edited and with an introduction by Luciano Formisano. New York: Marsiolio, 1992.

Morison, Samuel Eliot. *Admiral of the Ocean Sea: A Life of Christopher Columbus.* Boston: Little, Brown, and Company, 1942.

Schwartz, Seymour I. "The Greatest Misnomer on Planet Earth." *Proccedings of the American Philosophical Society* 146, no. 3 (2002): 264-81.

Zweig, Stefan. *Amerigo: A Comedy of Errors in History.* New York: Viking Press, 1942.

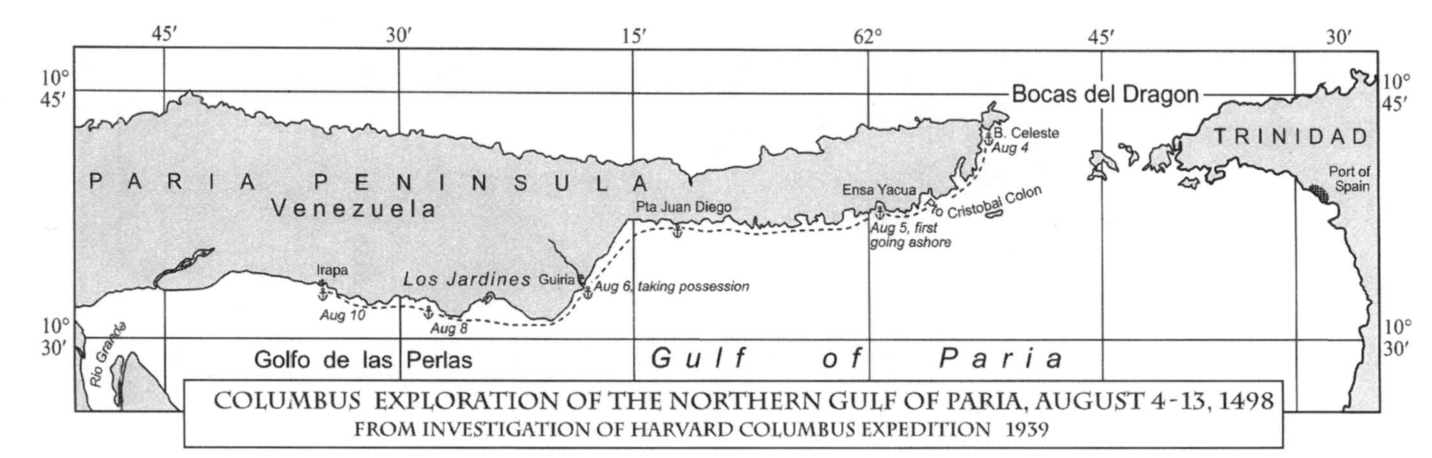

Fig. 1. "Columbus Exploration of the Northern Gulf of Paria, August 4-13, 1499." From Investigation of Harvard Columbus Expedition, 1939. After Samuel Eliot Morison, *Admiral of the Ocean Sea* (Boston: Little, Brown and Company, 1942).

Fig. 2a. Amerigo Vespucci (?1452 or 1454–1512). Engraving ascribed to Bronzini.

Fig. 2b. Statue of Amerigo Vespucci in the Uffizi Gallery, Florence, Italy.

Figs. 3a–3c. Four Voyages of Amerigo Vespucci detailed in *Lettera di Amerigo Vespucci delle isole nuovamente trovate in quattro suoi viagge* [Letter of Amerigo Vespucci Concerning the Isles Newly Discovered on His Four Voyages] (Florence, ca. 1505). Photograph courtesy of Luciano Formisano Marsilio, New York, 1992.

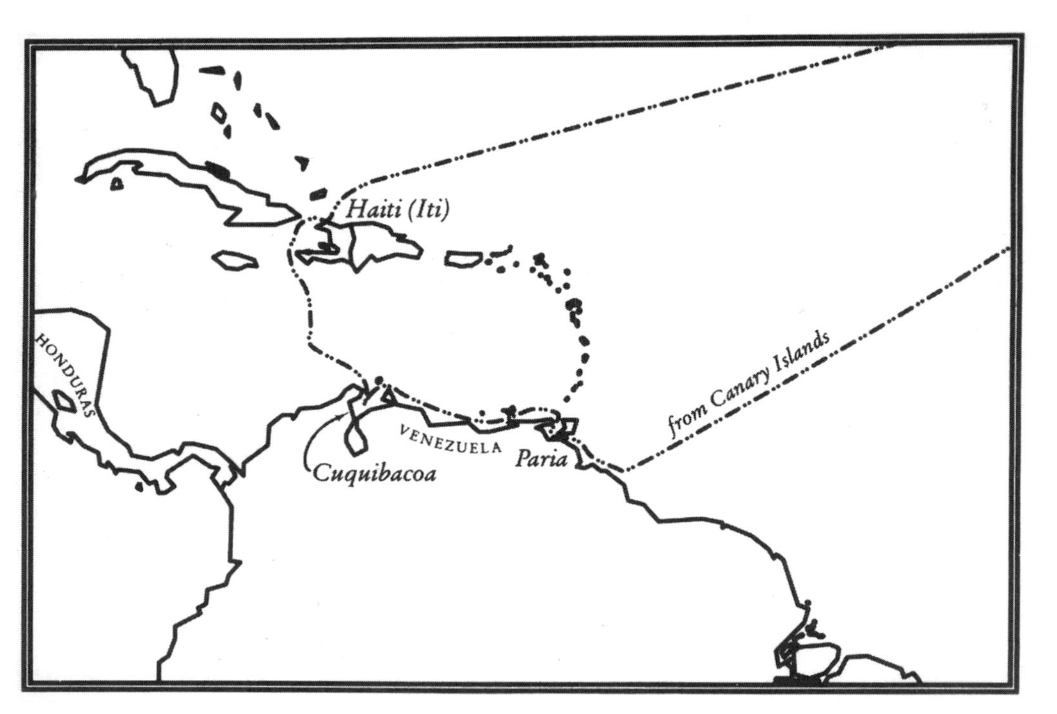

Fig. 3a. First Voyage: From Spain to Venezuela and Haiti. (Actually conducted in 1499).

Fig. 5. Magnified segment of Waldseemüller map (figure 4) containing the word "America" for the first time on a map.

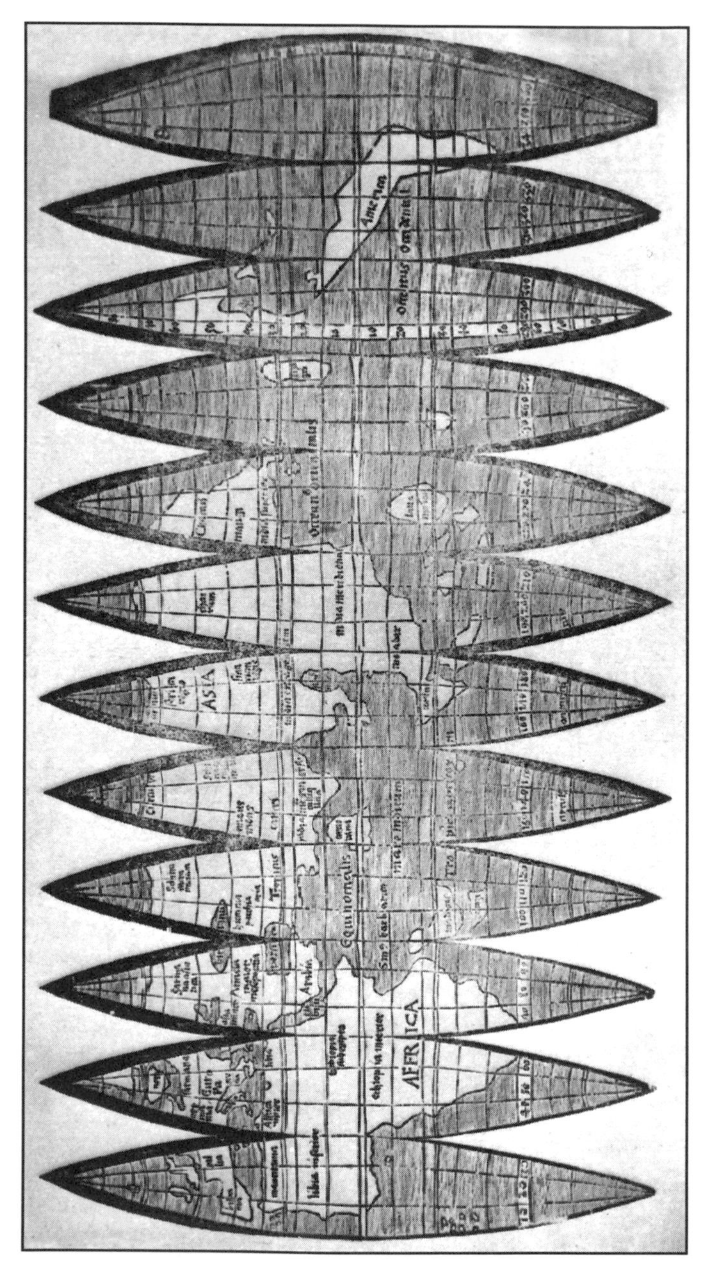

Fig. 6. Martin Waldseemüller. [No Title]. Set of 12 gores. St. Dié or Strasbourg 1507. Woodcut, 180 x 345 mm. Photograph courtesy of James Ford Bell Collection, University of Minnesota, Minneapolis, Minn. This set of gores for making into a globe was prepared as part of *Cosmographiae Introductio*.

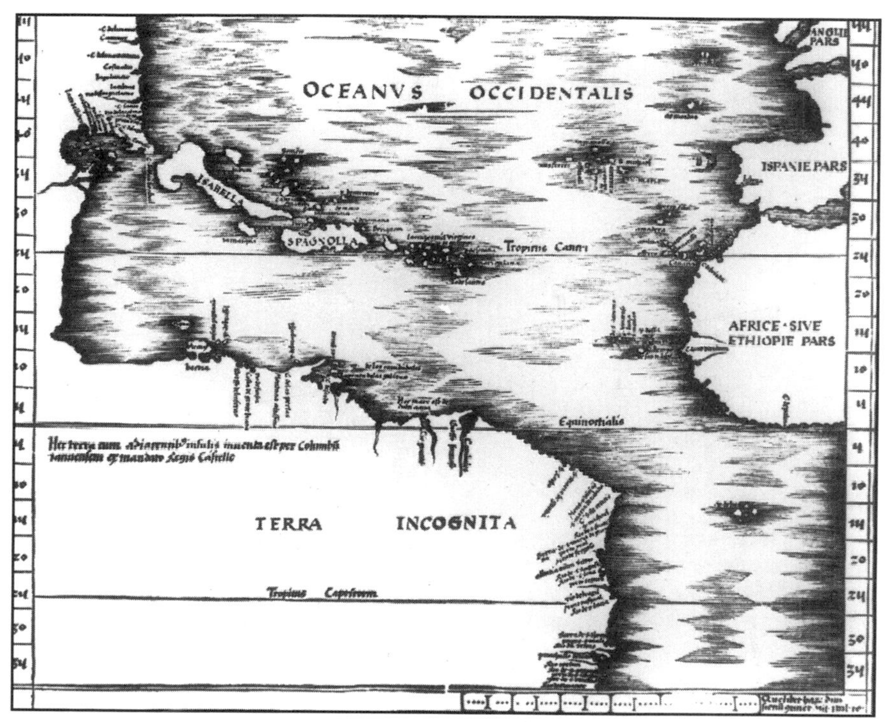

Fig. 7. Martin Waldseemüller. "TABVLA TERRE NOVE." Woodcut, 370 x 440 mm. From Ptolemy, *Geographia* (Strasbourg, 1513). Private Collection.

The map erases Vespucci's name from the southern continent in the New World. The land is named TERRA INCOGNITA, and the continent bears the inscription indicating that the land and adjacent islands were discovered by Columbus sent by authority of the King of Castile.

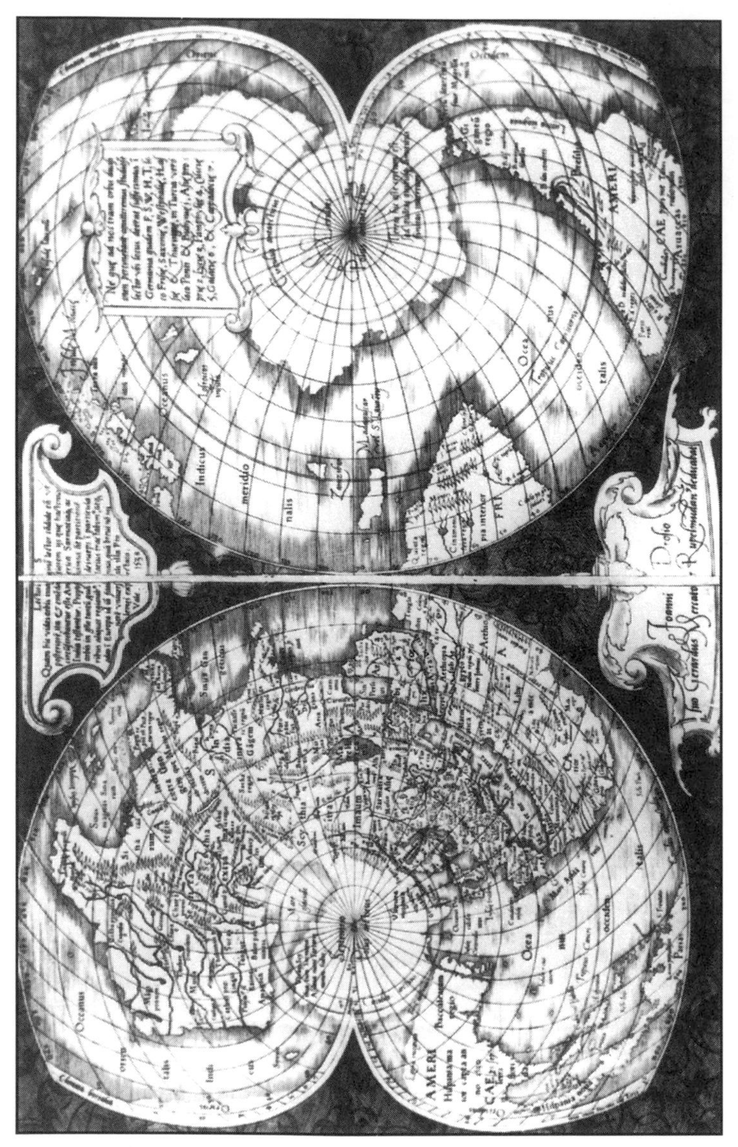

Fig. 8. Gerard Mercator. [No Title]. "Joanni Drosio suo Gerardus Mercator Rupelmudan dedicabat." Louvain, 1538. Engraving, 350 x 550 mm. Photograph courtesy of New York Public Library, New York.

The first world map compiled by Mercator and the first map to distinguish between, and to name, both North America (Americae pars Septentrionalis) and South America (Americae pars Meridionalis).

CHAPTER TWO

CINCHING A CORSET OF CONVENIENCE

Beginning in the sixteenth century and extending through over half of the seventeenth century, maps and narratives brought into focus the concept of a narrowing in the middle of the North American continent. A large inland sea, an extension of the Pacific Ocean or Sea of Cathay, indented the west coast of the continent, thereby allowing for the potential of a truncated overland passage between the Atlantic and Pacific Oceans. This geographic situation represented the next best circumstance, given that a convenient sea passage between the great oceans was not discovered. It would allow for the possibility of a shortened distance between western Europe and the Orient by combining sea voyages with a land caravan.

The sixteenth century began with the eyes of Europe focused westward to the setting sun in quest of a shorter route to the riches and spices of the Orient and East Indies, a route that would replace the long southerly passage and circumnavigation of the Cape of Good Hope at the southern tip of Africa. In the first decade of the sixteenth century, the Ionian and Aristotelian concept of the earth as a sphere with a great ocean encircling a single landmass had been replaced by textual and cartographic descriptions of continental land interposed between two oceans that separated western Europe from the Orient.

During the first quarter of the sixteenth century, explorers from Spain and Portugal continued to probe the Western Hemisphere within the constraints of the Papal Bull of Alexander VI and the 1494 Treaty of Tordesillas, which declared that newly discovered lands to the west of a north-south line that ran 370 leagues (roughly 1,175 miles) west of the Cape Verde Islands belonged to Spain, while Portugal was entitled to land east of that imaginary line. In 1500, Vincente Yáñez

Pinzón, a Spaniard, followed the South American coast to the mouth of the Amazon River. In April of the same year, Pedro Álvarez Cabral reached the coast of Brazil and claimed the land for Portugal. Also in 1500, Gaspar Corte Real crossed the Atlantic Ocean to Labrador, sailing under the Portugese flag, but was later lost at sea.

Spanish explorers dominated the activity during the rest of the first quarter of the sixteenth century. In 1501, Rodrigo de Bastidas landed on the east coast of Central America. Juan Ponce de León, the Spanish Governor of Puerto Rico, who had sailed on Columbus's second voyage, became the first European to set foot on land that would become part of the United States, when he explored Florida in search of the fountain of youth in the spring of 1513. He assigned the name "Florida" to the region because he landed during the Spanish feast of Easter, called "Pascua Florída." During that exploration, de León named a cape along the eastern shore "cabo de corrientes" (cape of currents). Shortly thereafter, the name was changed to "cabo canaveral" (cape canebrake) because of the canebrake reeds present along the shore. Cape Canaveral, called Cape Kennedy for a brief period, is the oldest persistent European name of a specific locale in the United States.

On September 25, 1513, Vasco Nuñez de Balboa stood on a peak in the Isthmus of Darien (current-day Panama) and viewed the Gulf of San Miguel , thereby becoming the first European to see a part of the Pacific Ocean from land in the Western Hemisphere. Within a year, the king of Spain was informed, and, in 1516, news of the discovery appeared in print in Peter Martyr's *De orbe nouo*. In 1519, the Spaniard Alonso de Pineda mapped the Gulf of Mexico and the Gulf Coast, demonstrating that a water passage from the Gulf to the Pacific Ocean did not exist. On Pineda's manuscript map, the word "Florida" appeared for the first time. Also in 1519, the Spanish conquistador, Hernán Cortés, established a base at Vera Cruz and went on to defeat Montezuma at Tenochtitlán (Mexico City) and subjugate the Aztec nation. Included in Cortés's report to the Spanish court in 1524 was a printed map of the old city and the Gulf of Mexico. On that map the word "Florida" first appeared in print.

In 1519, Ferdinand Magellan, a Portugese captain sailing under the flag of Spain, set out on a circumnavigation of the globe that ended when the ship, *Victoria*, arrived at San Lucar, Spain in 1522 without Magellan, who had been killed in the Philippines. During that voyage the Straits of Magellan were traversed and named, along with Tierra del Fuego.

Against this historical backdrop, the second major error in the mapping of America took place. The man responsible for the error that subsequently had historic consequences was Giovanni da Verrazzano, who, according to his biographer, Alessandro Bacchiani, was born of a wealthy and influential family, and was related to important personages in Florence.

But the issue of the birth of the explorer, Giovanni da Verrazzano, and his cartographer brother, Girlamo, has not been resolved. Until recently, it was held that Verrazzano was born in Florence or nearby Val de Greve, the family's ancient seat. His parents were thought to be Pietro Andrea di Bernardo da Verrazzano and Fiametta Capella, based solely on a statement of Vincenzo Coronelli in his *Atlante Veneto* of 1690. The engraved portrait of the explorer (see figure 9) by F. Allegrini, dated 1767, bears reference to this parentage. But neither Giovanni nor Girolamo appears among the recorded names of the children of the alleged parents.

In 1964, the historian and biographer, Jacques Habert, suggested that the Verrazzano brothers were actually born in Lyons, France, to parents who were part of a Florentine enclave living in that city as wealthy merchants and silk manufacturers. Records reveal that, in 1480, Giovanna, daughter of the international banker, Simon Gadagne, was married to Alessandro di Bartolommeo da Verrazzano. In support of this theory regarding Verrazzano's origin is the fact that Thomassin Gadagne, Giovanna's brother, was among those who sponsored the Verrazzano voyage of 1524.

The first mention of Verrazzano in any historic document relates to the suggestion that he accompanied Thomas Aubert of Dieppe on an expedition to Newfoundland in 1508. Although not substantiated, the occurrence would account for the statement made by Richard Hakluyt in his *Divers Voyages Touching the Discovery of America* of 1582, wherein it is written that when Verrazzano gave Henry VIII of England a map of his discoveries he said that he had already been "thrice upon these shores." Also, in Verrazzano's own journal of the momentous voyage of 1524, he wrote that, when he came upon Cape Breton and Newfoundland, he recognized the area from a previous expedition.

There is little in the way of evidence to determine Verrazzano's activities between 1508 and 1522, when his appearance in France was documented. A letter written by the Portugese viceroy of Mozambique, however, indicates that Verrazano was in Lisbon with Magellan in 1517 and traveled with him to Spain shortly before Magellan left on his voyage of circumnavigation. Verrazzano may also have spent a brief period in Cairo and other Middle Eastern cities.

Verrazzano's presence in France in 1522 is documented by a power of attorney that was given to him and recorded on November

14 of that year. In a chronicle written by a group of Portugese merchants in France to King Joâo of Portugal, it was reported that a Florentine, "Joâo Varezano" solicited endorsement from the king of France for a voyage of discovery in the New World.

The life of Verrazzano subsequent to his momentous voyage of 1524 is as difficult to define as his parentage. Because his patron, King François I of France, was imprisoned in Spain following his defeat at Pavia, Verrazzano sought patronage from Henry VIII of England and Joâo III of Portugal, but they both declined support. Shortly after François returned to his throne, an expedition was initiated in 1526–27 by Philippe Chabot, Admiral of France, and captained by Verrazzano. Although the ultimate goal of the expedition was the Spice Islands of the Moluccas, they did not reach that goal. The voyage, which extended over a year, proceeded from Honfleur, France, to the Amazon River and then along the eastern coast of South America to the Straits of Magellan. After failing to sail through the Straits, the ships turned eastward and continued to the Cape of Good Hope, but, unable to round the Cape, the expedition returned to France. Also as part of the Chabot enterprise, Verrazzano led an expedition to America in 1528, probably to the east coast of South America and the West Indies, but there is no documentation.

The circumstances of Verrazzano's death remain equally vague and unresolved. His contemporaries indicated that he was killed by Natives in 1528 or 1529 on a Caribbean island during his last voyage to America. Two hundred years later, Gonzalez de Barcia, in his book, *Ensayo cronologico para la historia general de la Florida,* implied that Giovanni da Verrazzano and the infamous French pirate, Jean Florin, were one and the same. The pirate was tried by the Spanish and hanged in 1527. The interpretation identifying Verrazzano with Florin is readily refuted by documents that attest to Verrazzano's activities subsequent to the date of the hanging.

The historic voyage of 1524, which, on one hand, explored the east coast of North America from northern Florida to Cape Breton for the first time, and, on the other hand, was responsible for a major misrepresentation of the geography of that continent, stemmed from Giovanni da Verrazzano's commission from François I of France to undertake a westward voyage in search of Cathay (Asia). Financial support was provided by Florentine bankers and merchants who resided in Lyons and were interested in the silk trade. The ship *Dauphine* and the salary of her captain were provided by the king. The captain was responsible for the ship's contents and personnel, while Verrazzano, as the master, was in command of the seamen and in charge of navigation and, therefore, deserving of the glory of discovery.

At the time of his departure in 1524, Verrazzano certainly knew that a large landmass separated the western coast of Europe from the eastern coast of Asia. As he stated in his chronicle, the Cèllere Codex (translated by Susan Tarrow for Lawrence C. Wroth's *The Voyages of Giovanni da Verrazzano*): "My intention on this voyage was to reach Cathay and the eastern coast of Asia, but I did not expect to find such an obstacle of new land as I have found: and if for some reason I did expect to find it, I estimated there would be some strait to get through to the Eastern Ocean." In part, Verrazzano's conclusions and anticipation were based on knowledge of the English discoveries in the northern region of North America that were well publicized in texts and on maps. In chronicling his voyage, Verrazzano wrote: "[W]e approached the lands which the Britons once found, which lies at 50 degrees."

The earliest printed reference to the English discoveries in the New World appeared in the second issue of Claudius Ptolemy's *Geographia*, which was published in 1508 in Rome. The work includes a world map by Johannes Ruysch that depicts Newfoundland and Labrador connected to Asia. In the text of the atlas, Marcus Beneventanus refers to a letter to the engraver of the map, in which Ruysch wrote that he himself had sailed from England to the Newfoundland area, probably on one of the fishing expeditions of a Bristol mercantile syndicate.

Verrazzano was also probably aware of the 1501 and 1502 voyages of Gaspar and Miguel Corte Real for Portugal. Although the two explorers failed to return from their individual voyages, ships from each exploration made it back to Portugal, and with them came word of the discovery of a new land—a land replete with timber and potential slaves, and bordering a sea teaming with fish. Evidence of the Corte Real discoveries appeared in the printed text of *Paesi nouamente retrouati,* published in 1507. In the Cèllere Codex, Verrazzano wrote: "We then left the land which the Lusitanians found, that is Bacalia [Newfoundland], so called after a fish." (*Baccalao* is Portugese for cod.) The Portugese discoveries were graphically depicted on several maps, including the Sylvanus world map of 1511, the first world map to be printed in two colors, black and red.

Printed books concerning the post-Columbian Spanish discoveries and explorations were also available to Verrazzano as he planned his voyage. In 1519, *Suma de geographia* by Martin Fernandez de Enciso was published in Seville. It described the northeastern coast of North America and sailing directions to the Caribbean islands and parts of Central and South America. It is also likely that Verrazzano knew of Magellan's discovery of a strait at the tip of the southern continent in the Western Hemisphere. This certainly would have reinforced Verrazzano's concept of a waterway traversing the North American continent, thereby reducing the sailing time from Europe to the Orient. Significantly, by 1524, news of the exploration of Yucatán, Mexico, and Florida had been widely disseminated throughout Europe.

The narratives of most voyages of exploration in the Western Hemisphere made in the fifteenth and sixteenth century have either been lost or are so vague as to require inferential deductions. By contrast, the chronicle of the voyage of 1524 under the command of Giovanni da Verrazzano, although interspersed with deduced inaccuracies by its author, provides the reader with a contemporary narrative by a literate and perceptive explorer. Although each of the three extant manuscripts chronicling Verrazzano's epic 1524 exploration is a transcription by a scribe of an original document, each transcription provides a contemporary narrative of the voyage.

One of the three manuscripts that were produced at the end of the sixteenth century includes nine pages of narrative reproducing Verrazzano's description of his voyage. It resides in the National Library of Florence and has been designated "Codex Magliabechiano, Miscellanea XIII, 89 (3)." The text was edited and published with an English translation by Joseph G. Cogswell in the *Collections of the New-York Historical Society* in 1841.

The second of the contemporary manuscripts, designated "MS. Ottoboniano 2202" in the Vatican Library, is a small folio in sixteenth-century hand. It bears the title "Copia di una lr̃a. di Giovanni da Verrazzano al Chr^{mo} Re Fran^{co} Re di Francia, della terra plui scopta in nomé di S. M^{ta}." (Copy of a letter of Giovanni da Verrazzano to King François of France . . . .) The manuscript presents a description of the Verrazzano expedition with only insignificant variations from the Codex Maglibechiano and the third contemporary manuscript, the Cèllere Codex.

The Cèllere Codex is, unquestionably, the most significant document related to Verrazzano's voyage of discovery. This manuscript is now part of a distinguished collection at The Pierpont Morgan Library in New York City. The document has its own romantic history. In 1909, a celebration commemorating the contributions of Henry Hudson and Robert Fulton to the development of New York City and the State of New York took place in New York City. That year, the Italian population of the city erected a large bust of Verrazzano in Battery Park, juxtaposed to the Lower Bay, Narrows, and New York Harbor, which he discovered.

Also in 1909, Alessandro Bacchiani, an Italian scholar, published a supposedly new version of Verrazzano's letter to François I, based on a manuscript that had recently been inherited by Count Giulio Macchi di Cèllere of Rome. Reconstructing the trail of the document written in 1524 to a 1909 residence in Rome is based both on facts and assumptions. The manuscript, consisting of twenty-four pages written by a scribe, but uniquely containing twenty-six annotations by Verrazzano himself, was given to Leonardo Tedaldi or Thomaso Sartini, merchants in Lyons, to be transported to Bonacorso Ruscellay, a banker in Rome whose partner was a relative of the explorer.

Twenty-seven years later, the document turned up in the possession of Paolo Giovio of Como, a highly regarded historian and antiquarian. In 1884, the document reappeared in the library of Count Alfredo de Szeth, a relative of Giovio. Count Alfredo de Szeth, the grandson of Francesco de Giovio, Paolo's kinsman, was the last male representative of that family. Francisco de Giovio's daughter, Donna Chiara, was the mother of Count Alfredo de Szeth. The slipcase of the Cèllere manuscript bears the inscription "Biblioteca Giovio-de-Szeth." In March 1911, through the intermediary of a gallery in Rome, the manuscript was purchased by J. Pierpont Morgan, and is known today as Morgan Ms.MA.776. It has thus returned to an area close to the geography featured in the explorer's narrative.

The Cèllere Codex, distinguished by the annotations in the margins and between lines, has assumed its nuclear position in the realm of scholarship related to Verrazzano's epic voyage of 1524. The document certainly established the authenticity of the voyage, a point that had been contested during the nineteenth century because no report was found in the royal archives. Shortly after his return to the port of Dieppe in France on July 8, 1524, Verrazzano drafted a communique to François I, detailing the journey. The document may have been sent while the king was occupied with other matters or in transit because of the death of Queen Claude at Blois on July 27. Not unlike Columbus's journal of his first voyage, Verrazzano's description failed to be included in any permanent archives.

The handwriting in the annotations in the Cèllere Codex differs from that of the text and is identical with the notes to Tedaldi and Sartini in Lyons and Ruscellay in Rome. Therefore, it has been concluded that the annotations are in Verrazzano's own script. Also, the descriptive nature of the annotations necessitated personal observations of a participant on the exploration. The annotations most probably did come from the hand of Verrazzano himself. We can picture the scenario in which the explorer structured an original draft in his native Italian language, and transcripts were made by a professional scribe to be used for revisions and translation into French for submission to the king. Using one of the copies of the original Italian document, Verrazzano added clarifying notes in his own hand for a communication to his Florentine supporters in Lyons and also for an ultimate revision for a French translation.

Contained within twelve leaves of paper, measuring 292 x 218 mm, with text on both sides, in the hand of a scribe, and twenty-six annotations in a contemporary hand (presumably Verrazzano's), is a detailed vivid description of the voyage that defined the east coast of North America from the northern portion of Florida to Cape Breton and Newfoundland. The Cèllere Codex begins with a report of Verrazzano's initial voyage with four ships that left France to explore new lands in the Western Hemisphere. A storm caused the loss of two vessels and forced the remaining ships, the *Dauphine* and the *Normanda,* back to Brittany. After a brief period of interfering with Spanish shipping off the Brittany coast, the *Dauphine* resumed its original mission and sailed to the Madeira islands.

On January 17, 1524, Verrazzano, aboard a single caravel, headed west on the voyage that would open the book of knowledge of the east coast of what would become the United States of America and, at the same time, instigate a misrepresentation of the geography of the North American continent. Having as his goal the discovery of a strait through the northern continent in the Western Hemisphere; with an appreciation that Spanish influence had been extended by Ponce de León to the Florida peninsula; and, perhaps, with the knowledge that Lucas Vasquez de Ayllón had already taken possession of the region as far north as 33° 30', Verrazzano chose a middle Atlantic transoceanic crossing. This course would avoid a conflict with the Spaniards but still offer him the opportunity of discovering a strait in the middle of the continent, or permit him access to the northern part of the continent, where it was thought to turn acutely to the west toward the Pacific Ocean.

Verrazzano was accompanied by a crew of fifty men, and the ship was equipped with arms and food provisions that would last eight months. According to the Codex, after fifty days of sailing westward along the 32nd parallel, during which the crew weathered a severe storm, landfall was made "in 34°" North Latitude. The ship came within a mile of low-lying land, where large fires were noted on the shore, indicating that it was inhabited.

The exact point where Verrazzano's first sighted land is unknown, but the explorer himself was convinced that it was located at 34°. This would place it at or near Cape Fear, North Carolina, which is several minutes south of that parallel. Verrazzano's "little book" that he referred to, in which he noted navigational procedures and tides in several regions during the journey, has never been found. It is more likely that the landfall was about nine minutes latitude south of Cape Fear, where the boundary line between current-day North and South Carolina reaches the ocean, a point more compatible with the description in the Cèllere Codex of land that stretched southward and that it contained no harbor for sixty miles.

After a brief southern exploration of the coast, the *Dauphine* was brought about and returned north, to the point at which land was originally sighted. Some of the crew went ashore in a small boat. The seashore was described as completely covered with fine deep sand, which rose as small hills (dunes) about fifty paces wide. Streams and inlets were interspersed along the shoreline. The coastal condition was good, with deep water and no dangerous rocks. Farther inland, there were elevated lands with fields and large forests containing a variety of trees. The region also contained an abundance of animals, including "stags, deer, and hares." Lakes and pools of running water and many different birds were noted.

The regional Natives were described as dark in color, of medium height, slightly taller than the crew members, with broad chests, strong legs and arms, and dark hair. Their well-proportioned bodies were naked, except for the loins that were covered with animal skins and furs. Some were ornamented with bird feathers. Several ethnologists regard this part of the Codex as the first detailed description of the Natives of North America.

The ship then proceeded in an easterly direction along the coast on a sea characterized by enormous waves. At one point, a sailor was sent swimming ashore with trinkets for the Natives. That "place" was the subject of an annotation (probably in Verrazzano's own hand) that was the genesis of the "false sea" concept.

The Italian script of the historic annotation in the Cèllere Codex, as transcribed by Frederick B. Adams Jr., reads:

> Appellavimus Annunciatam a die adventus, ovi trovasi uno isthmo de largheza de uno miglio e longo circa a 200 [insertion continued on verso of final leaf of manuscript] nel quale da la nave si vedea el mare orientale mezo tra occidente [changed from *l'oriente*] e septentrione. Quale è quello senza dubio che circuisce le extremita de la India, Cina e Catayo. Navicamo longo al detto isthmo con speranza continua di trovar qualche freto [a fine *de* crossed out after freto] o vero promontorio, al quale finisca la terra verso septentrione per poter penetrare a quelli felici liti del Catay. Al quale isthmo si pose da lo inventor Verazanio: cosci, come tutta la terra trovata, se chiamo Francesca per il nostro Francesco.

> We called it "Annunciata" from the day of arrival, and found there an isthmus one mile wide and about two hundred miles long, *in which we could see the eastern sea from the ship*, halfway between west [originally east] and north. This is doubtless the one which goes around the tip of India, China, and Cathay. We sailed along this isthmus, hoping all the time to find some strait ["to the end of" crossed out] or real promontory where the land might end to the north, and we could reach those blessed shores of Cathay. This isthmus was named by the discoverer "Varazanio," just as all the land we found was called "Francesca" after our Francis.

"Annunciata" is identified as the current-day Cape Lookout area, with its narrow sand barrier separating the Atlantic Ocean from the Pamlico and Albemarle Sounds. It is easy to picture Verrazzano standing on the deck of his ship and looking west. All that was visible a strip of land bounded on the west by a broad expanse of water, as far as the eye could see. Verrazzano had set out to find a strait that joined the Atlantic and Pacific Oceans in the middle of the North America continent. Absent the discovery of such a strait, the next best geographic situation would be a narrow isthmus (a mere one mile wide) between the two great oceans. Verrazzano's desire to discover a short route across the continental mass seemed to have been satisfied; his discovery might well have evoked a triumphal cry. It was this erroneous conclusion, the product of the need to satisfy a desire, that stimulated subsequent exploration and found expression on contemporary maps, confusing geographers and cartographers for over a century.

After the *Dauphine* and its crew left the region of "Annunciata," the journey continued along the coast in a northerly direction for about 160 miles, until they reached land that was even "much more beautiful and full of great forests." No mention is made of their having noted Chesapeake Bay. A period of over three quarters of a century would pass before that bay would be discovered and serve as the entry for the establishment of the Jamestown colony. After anchoring in a region deemed to contain no safe ports, twenty men went ashore, traveled about six miles inland, and spent three days exploring the region. The Native people of the land were described as being whiter than those who had been encountered earlier to the south. Their hunting and fishing were also described as were their canoes and the way they were constructed. According to a note in the margin, "we baptized [the area] 'Arcadia' on account of the beauty of the trees."

"Arcadia" was probably derived from the title of a very popular 1501 novel by Jacopo Sannazzaro. The author described the imaginary locale in his work as beauteous and characterized by many trees. It is generally held that Verrazzano assigned "Arcadia" to part of the eastern shore of Maryland, either in Worcester County, Maryland, or Accomac County, Virginia.

The voyage of the *Dauphine* continued in a northeasterly direction for slightly more than 350 miles. An annotation in the margin states: [translation] "We followed a coast which was very green and forested, but without harbors, and with some pleasant promontories and small rivers. We baptized it 'Costa di Lorenna' after the Cardinal: the first promontory 'Lanzone,' the second 'Bonivetto,' the largest river 'Vandoma,' and a little mountain by the sea, 'di S. Polo' after the Count." "Lanzome" probably referred to what is now Cape Henlopen, Maryland, while "Bonevetto" coincides with Cape May, New Jersey, and "Vandoma" indicates the mouth of the Delaware River. It is generally agreed that "S. Polo" was the name assigned to the Navesink Highlands south of Sandy Hook, New Jersey.

The explorers next came upon a wide river between two prominent hills, and took a small boat up-river to a densely populated land. At that point, the river widened to a "lake," which Natives crossed in small boats. An annotation in the margin of the text states: "called 'Angoleme' after the principality which you attained in days of lesser fortune; and the bay formed by this land we called 'Santa Margarita,' naming it after your sister, who surpasses all other matrons in modesty and intellect." King François I was Count of Angoulême; his sister Marguerite, Countess d'Alençon, was later to become Queen of Navarre. This small segment of the text of the Cèllere Codex raised the curtain on the history of the mouth of New York Harbor. It is certainly fitting that the modern bridge spanning the body of water bears the name of the explorer who discovered the site. Verrazzano's name has not been assigned to any other place along the entire east coast of North America. The names that Verrazzano bestowed on the region, that of the King of France and his talented sister, also have disappeared from modern maps.

Sailing continued eastward, always in sight of land, for about 250 miles, when the sailors discovered an island that "resembled the Aegean island of Rhodes" and named it "Alloysia" in honor of the King's mother. They sailed on another fifty miles and anchored in a safe harbor, where, during a fifteen-day stay, they encountered many Natives, whose appearance and activities were described in great detail in the text of the Codex along with the local plant and animal life. The text indicates that the land was located on a "parallel with Rome at

$40^{2/3}$ degrees." An annotation acknowledges the land "we called 'Refugio' on account of its beauty." It is impossible to assert the identity of the island in question, but the harbor referred to was certainly that of Newport, Rhode Island, in Narragansett Bay.

The *Dauphine* sailed from this harbor on May 6 and continued in a northeasterly direction along the coast for about 450 miles. An annotation indicates: "Within this distance we found sandbanks which stretch from the continent fifty leagues out to sea. Over them the water was never less than three feet deep; thus there is great danger sailing there. We crossed them with difficulty and called them 'Armellini.'" Ironically, Verrazzano gave the dangerous area the name of Cardinal Francesco Armellino, an intensely disliked papal official. The ship rounded Cape Cod and continued its voyage along the coast of Massachusetts and Maine. The mariners noted several off-shore islands; they named the largest group "Le III Figlie de Navarra" (The Three Sons of Navarre), which might have been Monhegan, Metini, and Matinicus—or alternatively Vinal Haven, Isle au Haut, and Swans Island. The ship ended its exploratory phase near "the land which the Britons once found, which lies in 50 degrees" and returned to France.

In the concluding segment of the Cèllere Codex, the author emphasized:

> Nevertheless, land has been found by modern man which was unknown to the ancients, another world with respect to the one they knew, which appears to be larger than our Europe, than Africa, and almost larger than Asia, if we estimate its size correctly. . . . All this land or New World which we have described above is joined together, but is not linked with Asia or Africa (we know this for certain), but could be joined to Europe by Norway or Russia; this would be false according to the ancients, who declare that almost all the north has been navigated from the promontory of the Cimbri to the Orient, and affirmed that they went around as far as the Caspian Sea itself. Therefore the continent would lie between two seas, to the east and west; but these two seas do not in fact surround either of the two continents . . . .

The document ends with the statement: "In the ship 'Dauphine' on the VIII day of July. M.D.XXIIII. Humble Servant Janus Verazanus."

Giovanni da Verrazzano must be credited with the seminal definition of the east coast of the continent from north of the Florida peninsula to Newfoundland. And he was the first to identify what would become New York Harbor and Narragansett Bay. He failed, however, in his primary quest for a short water route from Europe to Cathay and a strait through North America that would join the Atlantic and Pacific Oceans. But he believed that he had discovered a combination of a more direct transoceanic passage and an easy overland journey. Although a compromise, it would have been preferable to passage through the Straits of Magellan. One could imagine a fleet of merchant ships sailing from Europe to the east coast of North America and the transfer of goods to an overland caravan, which would, over a short distance, transport the cargo to vessels on the west coast of the narrow peninsula. These ships would then proceed to the Orient. The riches and spices of the Orient would follow the same route in the opposite direction.

Thus, the seed was sown in a brief handwritten annotation, found in a single manuscript, known as the Cèllere Codex—the seed that would serve as the genesis of a new idea: the declaration that located in the midst of the North American continent a point where the land was contracted, thereby allowing easy access from the eastern shore to the Pacific Ocean. The situation was not unlike that which pertained to the misnaming of "America," in which case two printed documents, one consisting of four pages and the other of thirty-two pages, led to the definitive placement of Amerigo Vespucci's name on a map that included the New World.

For the "False Sea" of Verrazzano to be better appreciated and accepted, a more dramatic statement was required. As was the case in the naming of America, that statement was made on a contemporary map, and subsequent maps offered strong pictorial evidence that would perpetuate the concept for over a century. The cartographic misrepresentation of Verrazzano's concept of a narrow strip of land in North America separating the Atlantic and Pacific Oceans had to be a modification that was superimposed on the geography depicted on antecedent maps related to the New World. Generally, these maps were also graphic interpretations of verbal and written statements.

Henry Harrisse, the renowned French scholar of history and cartography, in the first great book on the mapping of the New World, lists several maps of America believed to have been executed before 1500, but not one of these has been uncovered. Doubtless this was, in part, related to the fact that every country possessively attempted to prevent the spread of recently discovered geographic information in order to maintain a power base in the New World. Thus, the Juan de la Cosa portolan world chart, a pen and ink and watercolor manuscript on oxhide in the Museo Naval in Madrid, is the oldest extant map to depict part of the New World (see figure 10). Bearing the date of 1500, the chart not only incorporates pictorial representations of

Columbus's discoveries in the Caribbean Sea and names Cuba, but also, for the first time, vividly depicts a segment of the North American continent. Placed along the northeastern coastline of America are five English standards. The phrases "mar descubierta por yngleses" (sea discovered by the English) and "cauo de ynglaterra" (cape of England) appear and provide evidence of John Cabot's 1497 voyage of discovery. The North American continent contains twenty inscriptions, including seven capes, a river, an island, and a lake. The names are not significant, nor are they found on any subsequent maps. The Juan de la Cosa map includes the current Labrador-Newfoundland regions, and, perhaps, even the Gulf of Maine, the Cape Cod peninsula, and New York's lower bay.

The "Cantino Planisphere," located in the Biblioteca Estense, Modena, Italy, is a manuscript drawn with pen and ink and color on vellum and bears the date of 1502; it pictures Columbus's discoveries and part of South America, and it offers the suggestion of the Florida peninsula. By positioning Newfoundland as an island east (therefore on the Portugese side) of the papal line of demarcation, it reflects the Portugese viewpoint and ignores the English discoveries. The land representing Newfoundland and Labrador is named "Terra del Rey del Portuguall."

In 1506, the oldest extant printed map that included part of North America came off a press in Venice or Florence. The map was designed by Giovanni Matteo Contarini and engraved by Francesco Rosselli. It was discovered in 1922 and is now located in the British Library in London. The coniform or fan-shaped copperplate represents an intermediary stage between the old Ptolemaic system and modern world geography. On the map, in the New World, a wide sea separates Columbus's discoveries from Labrador and Newfoundland, which are connected to Cathay (Asia). Omission of a coastline to the west of Cuba is evidence that the cartographer accepted the belief that Columbus had reached Asia. Added evidence is found in the inscription off the coast of Asia stating that Columbus sailed to Ciamba (the Orient). The map also reflects the discoveries of Cabot and Corte Real. An inscription referring to the North American continent reads: "Hanc terram invenere naute Luisitanor[um] Regis" (The seamen of the King of Portugal discovered this land).

The map by Johann Ruysch, printed for Ptolemy's *Geographia* in 1507–8 in Rome, depicts the world on a similar coniform projection with Newfoundland, named "Terra Nova," as an extension of Asia. Offshore islands bear the name *Baccalauras*, an early reference to the codfish in that area. The South American continent, for the first time, is named *Mundus Novus*, the term ascribed to Amerigo Vespucci, whose modified Christian name, America, was permanently imprinted on maps beginning with Waldseemüller's production of a large plane (flat) map of the world and a globe gore in 1507. Those maps, their origins, their implications, and their effects are discussed in detail in the first chapter.

In 1511, Peter Martyr, the Milanese humanist and tutor of the children of King Ferdinand and Queen Isabella of Spain, produced the first map printed in Spain emphasizing Columbus's discoveries. On that map, there is a suggested southern coast in the region of what would be defined two years later by Ponce de Leon as Florida. The 1511 world map of Bernard Sylvanus, published in Venice, with its red lettering, shows no evidence of a North American landmass other than an island, "terra laboratorus." Also in 1511, Vesconte de Maggiolo, who would later play a critical role regarding the Verrazzanian concept, produced a manuscript map on which are inscribed the legends "Terra de Laueradore de rey de portugall" and "Terra de los Ingresy." This map was the earliest Italian delineation of the northern part of the New World.

In 1512, a woodcut mappemonde (world map) by Jan ze Stobnicy (Joannes de Stobnicza) was published in Crakow. The one extant copy of the map is located in the Österreichische Nationalbibliothek in Vienna. The map presents a continuous coastline between North and South America, the two continents joined by a relatively narrow landmass. This map is considered to be a plagiarism of the 1507 Waldseemüller large plane map. In 1513, Martin Waldseemüller incorporated two new maps in the Strasbourg edition of Ptolemy's *Geographia*. The world map, sometimes known as "the Admiral's map," referring to Columbus or Vespucci, includes none of the North American continent. By contrast, the other map, "Tabvla Terre Nove," offers a depiction of a Florida peninsula, a southeast coast, and what could be interpreted to be the Gulf Coast. The same representation appeared on Gregor Reisch's "Typus Universalis Terre Iuxta Modernorum Distinctionem et Extensionem per Regna et Provincias" in the 1515 edition of his *Margarita philosophica nova* and on Waldseemüller's "Carta Marina Navigatoria Portugallen Navigationes atque Tocius Cogniti Orbis," published in Strasbourg in 1516. The latter is, to that date, the most accurate nautical chart of the modern world. The so-called "Paris" or "Green Globe," ascribed to 1515 and located in the Bibliothèque Nationale de France in Paris, presents the word "America" in four places, including, for the first time, on North America.

Peter Apian's map, published in Vienna in 1520, is similar to the Reisch and Waldseemüller representations of North America, showing a southeast coast, a Florida peninsula, and a Gulf Coast. The manuscript map of the Pineda expedition, drawn about 1519, was the earliest map to outline correctly the Gulf of Mexico and to demonstrate the absence of a passage from the Gulf to the Pacific. This manuscript was followed by the 1524 printed map by Hernán Cortés, a woodcut, which was included in Cortés's *Praeclara Ferdina[n]di Cortesii de noua maris oceani Hyspania narratio*. It is the first printed map to name Florida and the first relatively accurate printed map of the Gulf Coast.

How many of these few extant maps or other unknown maps were available to those mapmakers responsible for transforming the words of Verrazzano into a graphic representation is purely conjectural. Surely the number was very small. The task of the cartographers was to alter the representation of the east coast of North America from a previously absent or speculative entity by translating the often vague descriptions of the explorer into a meaningful and accurate graphic presentation, a daunting challenge in a time when both the navigators' and cartographers' tools were limited and primitive.

Five maps drawn between 1525 and 1529 offer a pictorial representation of the geographic text of the author of the Cèllere Codex or an equivalent report. As such, they are regarded to be primary maps from which all later maps extending to 1651 were derived.

The first of the primary maps has never been seen by modern scholars. Known as the Verrazzano–Henry VIII Map of 1525 was a manuscript that Verrazzano allegedly presented to the King of England in 1525 or 1526. It is unknown whether the explorer personally gave the map to the King or whether Verrazzano even went to England himself seeking support for a second voyage to America. The preface to Richard Hakluyt's *Divers Voyages* of 1582 reports: "Secondly that master John Verazzanus, which had been thrise on that coast, in an olde excellant mappe which he gave to King Henrie the eight, and is yet in the custodie of Master Locke, doth so lay it out, as it is to be seene in the mappe annexed to the end of this boke, being made according to Verazzanus plat." This is offered as evidence for the existence of the map.

Reinforcement of the map's presence appeared in Hakluyt's *Discourse Concerning Western Planting* of 1584 in which he supported the existence of the Sea of Verrazzano that expedited passage to Cathay. Hakluyt was quite specific when he wrote: "mightie large olde mappe in parchment made as yt should seeme by Verasanus, traced all along the coaste from fflorida to Cape Briton with many Italian names, wch laieth oute the sea making a little necke of lande in 40 degrees of latitude much like the streyte necke or Istmus of Dariena." The 40 degrees referred to by Hakluyt was an error that placed the "False Sea" too far north. This error was also present on other maps. It is than more likely that the Verrazzano–Henry VIII map was drawn by Girolamo da Verrazzano, Giovanni's brother and an accepted cartographer, because the 40° location appears on the extant map that he is known to have drawn. It has also been suggested that the Verrazzano-Henry VIII map prompted the king to dispatch John Rut with two vessels to the Orient. That voyage's chronicles are missing, but Spanish reports indicate that two ships left England on June 10, 1527, and reached Newfoundland on July 21. One of the ships was sighted in the Caribbean in November of that year.

The oldest extant map to incorporate information gained from Giovanni da Verrazzano's exploration is a large mappemonde (see figure 11) by Juan Vespucci, Amerigo's nephew, dated 1526. The map depicts, for the first time, the explorations north of the Florida peninsula conducted in 1521 and 1525 under the sponsorship of Lucas Vásquez de Ayllón of Santa Domingo. The map also manifests knowledge of Verrazzano's voyage, as evidenced by the appearance of a river at about 34° North Latitude bearing the name, "R. da Sa Terazanas." It is possible that the copyist substituted a "T" for a "V." The use of "T" as the initial letter of Verrazzano's name was also used in a letter written in 1527, one year after the map was drawn. The proximity of the named river to Verrazzano's landfall adds to the probability that the cartographer was including information from Verrazzano's voyage. There is no suggestion of a body of water indenting the west coast of North America and a consequent narrow isthmus separating the Atlantic and Pacific Oceans.

The first map to depict a False Sea of Verrazzano creating a narrow isthmus in the North American continent was drawn by Vesconte de Maggiolo (Maiolo), a Genoese cartographer of note. The Biblioteca Ambrosiana of Milan, where the map resided, was destroyed during the bombing of Milan in World War II. The map assumed primacy as a map incorporating Verrazzano's conclusions when the map's correct date was proven to be 1527 and not its apparent date of 1587. The "2" had been changed to an "8." Fortunately, a full size photograph was made in 1903, allowing for study of this important historical document (see figure 12).

The map was drawn three years after Verrazzano's memorable voyage. During that interval, three other exploratory journeys took place. In 1525, Lucas Vásquez de Ayllón, a wealthy lawyer and judge residing in Hispaniola, sent Pedro de Quexos on a voyage that extended north as far as the Outer Banks of North Carolina. In 1526, Ayllón himself landed in the vicinity of the Cape Fear River, and moved south to the Rio Gualdape (Guadalupe) near what is currently Wilmington, North Carolina. He established a settlement there, but it was abandoned in less than two years, The only remnant of the settlement was the name Chicora, derived from a local Indian whom Quexos had enslaved during an earlier trip. The Indian learned Spanish, was baptized, and accompanied Ayllón to Spain in 1523, when Ayllón pleaded for a patent of possession for his discoveries. For many years, "Chicora" was the name applied to the coastal region between Charleston Harbor and the Cape Fear River.

Also, shortly after Verrazzano's voyage, Estévan Gomez, a Portugese mariner who had deserted Magellan during the circumnavigation of 1521, explored the east coast of North America. Sailing from Cuba in 1525, Gomez bypassed the coast between New York Bay and Cape Cod, and voyaged from New England to Nova Scotia in search of a northern passage to the Orient. There is no written record of the exploration, but the coastal findings were soon related to cartographers and were initially depicted on an 1525 manuscript map, known as the "Castiglione World Map." From that point in time, maps presented the east coast of the North American continent in accordance with either the description of Verrazzano or that of Gomez.

The Maggiolo map presents latitudes which are consistently too low. The tip of Florida is shown at 20° rather that the correct 25° North Latitude. But the general representation of the coast and topography is more in concert with the facts. The Verrazzanian isthmus and "False Sea" are not described on the map, but are well delineated, and at the isthmus's southernmost point, the name "la nuntiata" appears. This is surely the equivalent of "Annunciata" in the Cèllere Codex. Maggiolo, like Verrazzano, ascribed the name "Fracesca" to the landmass to the north of the isthmus. Many of the place names assigned by Verrazzano in the marginal notes in the Cèllere Codex appear on the map, suggesting that either a revised Codex or a sketch map by the explorer was available to the cartographer. The Maggiolo map certainly provides a graphic record of the first continuous exploration of the North American coast between upper Florida and Newfoundland.

The map by Girolamo da Verrazzano, the explorer's brother, accomplished the same effect (see figure 13). The original is in the Vatican Library, and has been assigned a date of 1529, based on an inscription that denotes the voyage made five years before. In contrast to the Maggiolo map, the Verrazzano map places the tip of Florida at 33° 8' North Latitude, too far north. Girolamo da Verrazzano's map depicts the Verrazzanian isthmus and "False Sea," and contains a legend indicating that the isthmus is six miles wide, different from the

Cèllere Codex annotation of one mile wide. The name "Annunciata" appears north of the isthmus. The land north of the isthmus is named "Nova Gallia sive Ivcatanet." An inscription in the region reads: "Verrazana sive nova Gallia." Many of the place names designated in the Codex appear on the map.

Another map ascribed to Girolamo da Verrazzano has been added to the group of primary cartographic representations of the 1524 expedition. It is a large world map, which was found in the second half of the twentieth century in the National Maritime Museum at Greenwich, England. No cartographer's name appears on the map but, because the American east coast is depicted with the same geography and place names as Girolamo da Verrazzano's 1529 map, and because of the handwriting, the work has been accepted as having been executed by the explorer's brother. It was probably initially produced not long after 1528, but contains additions from ca. 1540–42.

Graphic representation of the "False Sea" and Atlantic coast line described by Giovanni da Verrazzano persists on cartographic works that were derived from the five primary maps. Because Verrazzano sailed under the flag of France, logically his findings found expression in French cartography. The most notable example is a copper globe made by Robert de Bailly and dated 1530. Three copies are known to exist, one in The Pierpont Morgan Library (see figure 14) accompanied by its gores on a flat surface and complimenting the Cèllere Codex; one in the Bibliothèque Nationale de France, Paris; and one in the Museo Lazar Galdiano, Madrid. The depiction of North America on the globe derives from the Girolamo da Verrazzano map; the country north of the Verrazzanian isthmus is named "Verrazana." The "False Sea" and coastline also follow that of Girolamo's depiction, as does the 40° placement of the southern end of the "False Sea." At the same time, the Paris publications of the 1531 double cordiform (heart-shaped) and the 1534 single cordiform maps, both by Oronce Finé, present the coastline described by Verrazzano, but with North America connected to Cathay and without a "False Sea."

Another exploration of consequence to the North American continent occurred in 1534, when Jacques Cartier sailed with two ships from St. Malo under the French flag. After making landfall at Cape Bonavista in Newfoundland, the ships eventually made their way to the northern entrance of the Gulf of St. Lawrence. The ships continued to current-day New Brunswick, and then sailed along the Gaspé Peninsula before returning home. On Cartier's second voyage in 1535, he named La Baye Sainct Laurins for the saint whose festival fell on the next day, and the vessels then entered the river, which was subsequently assigned the same name. Cartier applied the name

"Canada," an Indian word meaning "a collection of homes," to the region between current-day Quebec (later named by Champlain) and the Saguenay River. The ships continued upriver to the Indian village of Hochelaga, where Cartier named the overlooking adjacent hill "Mont Real" (royal mountain). Cartier's voyages had no effect on the concepts of Verrazzanian geography.

Giovanni da Verrazzano's geography continued to be depicted on maps throughout the first half of the sixteenth century. Between 1535 and 1552, Battista Agnese, a Genoese cartographer working in Venice, compiled several beautiful manuscript atlases. In his "Agnese World Map," the east coast of North America is based on the Gomez voyage rather than on the findings of Verrazzano. By contrast, both Agnese's Oval World Map and his Atlantic Hemisphere map depict Verrazzanian coastal concepts, including a continuous coastline from Florida to Labrador, a coastline from New York Bay to Cape Cod, and a "False Sea" (placed too far north) (see figure 15). Between 1535 and 1552, Agnese's atlases maintained the representation of the Verrazzanian Sea. Over the years, Agnese offered several variations on Verrazzano's theme.

On Agnese's Oval World Map of ca. 1538, there is a strait in northeast North America. Within that strait is a line of dots running from the coast of France to Cathay. In a later map, the strait was abandoned, but a route contains the legend, "el viaza de fransa," crossing the ocean from Franzia to the Verrazzanian isthmus and on to Cathay via the Verrazzanian Sea. In a later work, the route with the same name stops in the middle of the North American continent and does not extend to the Orient. In Agnese's latest works, the route and legend disappear but the isthmus and "False Sea" persist.

Between 1540 and 1578, the concept of the Verrazzanian Sea continued on the maps of Sebastian Münster. Münster's world and American maps first appeared in an edition of Ptolemy's *Geographia*, published in Basel in 1540. The map entitled "Novae Insulae, XVII Nova Tabvla" is the first printed map (see figure 16) that clearly depicts the New World as a distinct single insular landmass. It also presents the Strait of Magellan and Magellan's ship, the *Victoria*. The map's long life span persisted with its 1544 publication in Münster's *Cosmographia universale*, followed by thirty-two editions. In each instance, the map depicted a "Sea of Verrazzano."

The 1552 manuscript world map by Georgius Calapoda and its engraved copy, known as the "Florentine Goldsmith's Map" (see figure 17), presents both the Verrazzanian coast and sea. The lands to the north of the isthmus are named "Tierra del Bachalaos" and "Tierra del Labrador," and are connected to the Asian landmass.

During the second half of the sixteenth century, the sea, which indented the west coast of the North American continent, found its expression in English texts and on English maps. In Richard Hakluyt's 1582 *Divers Voyages,* the author indicates that a "Master John Verasanus" gave a map to Henry VIII, and that map was in the possession of Michael Lok, worldwide traveler, entrepreneur, and Governor of the Cathay Company. The book contains a map by Michael Lok depicting the Verrazzanian isthmus and sea, which bears the legend "Mare de Verrazzana 1524." The southern end of the sea is located at 40°, in keeping with the Maggiolo manuscript map. On the expanse of land to the north are imprinted: "Montes S. Johannis," "Saguenay," "Hochelaga," "Canada," "Cortereal," and "R. ELIZABETH." The land south of the isthmus is named "Apalchen" (see figure 18). The map was offered deliberately as evidence in support of a Northwest Passage to Asia.

Also in 1582, a manuscript circumpolar map made by John Dee, a strong advocate of a Northwest Passage, for Sir Humphrey Gilbert, included a "Mare Verazana 1524." The inscription appears in a narrow body of water that extends west from an isthmus, located at about 34° on the east coast of North America, to the Pacific south of the Lower California peninsula.

England's explorations in the New World essentially ceased after John Cabot's 1497 voyage until the reign of Elizabeth I. During her reign, England's interests initially focused on attempts to discover a more direct route to the northern part of Asia, a Northwest Passage. The three voyages of Sir Martin Frobisher, conducted between 1576 and 1578 and the three trips of John Davis in 1585, 1586, and 1587 stemmed from the quest for a Northwest Passage. In 1578, Sir Humphrey Gilbert received from Queen Elizabeth letters patent for land in North America. In 1583, Gilbert, himself, set out to establish a settlement but his ship sank in a storm. The above-described 1582 circumpolar manuscript map by John Dee could have been used as propaganda for that voyage.

Gilbert's grant was transferred to his half-brother, Walter Raleigh, who, in 1584, sent Captain Philip Amadas and Arthur Barlowe to search out a site for colonization. Upon reaching North America, they entered an inlet at the northern end of Hatarask Island (the Indian name for a barrier island that extends south to Cape Hatteras), and eventually came upon another island that the Indians called Roanoke, meaning in Algonquian "white shell place." After six weeks, they returned to England with glowing comments to encourage colonization. Sir Walter Raleigh, who enthusiastically moved to establish such a colony, had been knighted and received royal allowance from Queen

Elizabeth (known as the Virgin Queen) to name the colony in her honor "Virginia." "Virginia" was the first English name placed on American soil.

In 1585, Sir Richard Grenville sailed from Plymouth with seven ships, and after exploring the Carolina sounds, left Ralph Lane as Governor and 108 colonists. Included among those who returned to England was the watercolorist John White. The colony was beset by antagonism within their group and hostility from local Indians. In the summer of 1586, the remaining colonists returned to England with Sir Francis Drake. Later that year, Grenville returned to the region and, finding the colonists gone, left a holding force of 15 men; all perished.

John White's "La Virgenia Pars," a manuscript map executed with pen and ink and watercolor is located in the British Library in London. This beautiful map shows the southeastern coast from Chesapeake Bay to the tip of Florida (see figure 19). A channel runs from Port Royal at 34°North Latitude to a large body of water, thereby suggesting a passage to the Pacific Ocean similar to the Sea of Verrazzano.

In 1587, Raleigh dispatched a group of 180 persons under the leadership of John White as Governor to establish, according to Raleigh's charter, the "Cittie of Raleigh in Virginea," with specific instructions to locate on Chesapeake Bay. The sailing master, however, put the colonists ashore at Roanoke. The settlers included White's married daughter, who, in August 1587, gave birth to Virginia Dare, the first child of English parents born in America. White returned to England to establish a better supply line and did not return to the colony until the summer of 1590 at which time he found no survivors. This unfortunate settlement became known as the Lost Colony.

In the early part of the seventeenth century, the English finally established a firm foothold on the North American continent, in Virginia, facilitated by a 1606 treaty between James I of England and the King of Spain allowing England to devote substantial resources toward American colonization. Two companies were chartered by James I in 1606 to found colonies in America: the Virginia Company of London (London Company), which was granted permission to plant a colony 100 miles square between 34° and 41° North Latitude, and the Virginia Company of Plymouth (Plymouth Company), which was allowed to found a colony between 38° and 45° North Latitude. The overlapping area was open to settlement by either company, although in order to provide a buffer, neither company could settle within 100 miles of the other.

The London Company underwrote a venture that began on December 20, 1606, when three ships, *Susan Constant*, *Godspeed*, and *Discovery*, under the command of Captain Christopher Newport left London. On April 26, 1607, the group reached Chesapeake Bay, and, on May 14, after preliminary exploration of the bay, selected a low swampy peninsula as the first site for settlement, naming it Jamestown, after the King. John Smith explored Chesapeake Bay, in part, in search for Roanoke survivors, and also in quest of a water passage to the west; he described the geography in his 1608 *A True Relation of Such Occurrences and Accidents of Noate as Hath Hapned in Virginia.*

The mother map of the region, John Smith's "Virginia," was initially issued as a separate publication in 1612, and was subsequently included in numerous editions of Smith's *Generall Historie of Virginia, New-England, and the Summer Isles,* published in London from 1624 to 1632. In the 1625 publication *Purchas his pilgrimes*, the esteemed mathematician Henry Briggs provided a "Treatise of the North-West Passage to the South Sea, through the Continent of Virginia, and by Fretum Hudson," expressing the "hope that the South Sea may easily be discovered over land," citing the "constant reports of the Savages . . . of a large Sea to the Westwards." During the remainder of the first half of the seventeenth century, maps published in Leyden by Joannes de Laet, and in Amsterdam by Willem Janszoon Blaeu, by Henricus Hondius, and Jan Jansson, all depicted the eastern part of Virginia devoid of any of Verrazzano's concepts.

But the concept of a commercially advantageous narrow isthmus in the region of Virginia or Carolina was slow to die. Over a century and a quarter after Verrazzano made his misinterpretation, John Farrer of London produced a map in 1651 that strongly endorsed, with modification, Verrazzano's written findings. The map (see figure 20) bears a legend imprinted below "The Sea of China and the Indies." It states:

> Sir Francis Drake was on this sea and landed Ano. 1577 in 37 deg. where hee tooke Possession in the name of Q. Eliza: Calling it New Albion. Whose happy shoers (in ten dayes march with 50 foote and 30 horsemen from the head of Jeames River, over those hills and through the rich adjacent Vallyes beautyfied with as proffitable rivers, which necessarily must run into y^{r}. peacefull Indian Sea) may be discovered to the exceeding benefit of Great Britain and joye of all true English.

Verrazzano's one-mile-wide isthmus had been expanded to one that required ten days for crossing. On the map, Virginia extends from the Atlantic to the Sea of China. Also shown on the map is a narrow isthmus that separates the upper "Hudsons River" from "A mighty great Lake" that connects with the Sea of China. Thus, a second expeditious route to the Orient was added, making the geography more attractive.

After the dramatic graphic representation and textual statement of the Farrer map, the Verrazzanian isthmus and "False Sea" would remain permanently deleted from future maps. But the dream persisted for a while longer. *The Discoveries of John Lederer,* translated from the Latin by William Talbot and published in London in 1672, provides an epilogue. John Lederer explored the Blue Ridge Mountains and the Piedmont during three journeys in 1669–70 and recorded his travels. In the text, the author wrote:

> They are certainly in great errour, who imagine that the Continent of North-*America* is but eight or ten days journey over from the *Atlantick* to the *Indian* Ocean. . . . Nevertheless, by what I gather from the stranger Indians at *Akenatsy* of their Voyage by Sea to the very Mountains from a far distant Northwest Country, I am brought over to their opinion who think that the Indian Ocean does stretch an Arm or Bay from *California* into the Continent as far as the *Apalataean* Mountains, answerable to the Gulfs of *Florida* and *Mexico* on this side. Yet I am far from believing with some, that such great and Navigable Rivers are to be found on the other side [of] the *Apalataeans* falling into the Indian Ocean, as those which run from them Eastward.

In the introduction, Talbot concluded, addressing Anthony Lord Ashley-Cooper, later Earl of Shaftesbury and one of the Lords Proprietors of Carolina, that the arm of the Pacific Ocean extending toward the east was located in the region of the Province of Carolina: "From this discourse it is clear that the long looked-for discovery of the *Indian* Sea does nearly approach: and *Carolina*, out of her happy experience of your Lordships success in great undertaking, presumes that the accomplishment of this glorious Designe is reserved for her."

The misinterpretation of the great explorer, Giovanni da Verrazzano, was, in part, an expression of his cherished hope. The idea of an area of a narrow isthmus within North America presented itself as a convenient solution for Verrazzano. For the explorer who made history's most monumental voyage along the east coast of the continent, a narrow strip of land between the Atlantic and Pacific Oceans represented a convenient compromise. Having failed to discover the desired waterway through the middle of the continent that would expedite commerce between Europe and the Orient, how pleased he must have been to stand on the eastern shore of the middle of North America, look to the west, and view a broad expanse of water, which he interpreted to be the Sea of Cathay. It was next best alternative to a midcontinental strait.

It was logical for Verrazzano to include in his narrative what he regarded to be his major discovery. It was also logical for this important "discovery" to become rapidly incorporated on maps. Because of its appeal, the misrepresentation would persist on maps for over seventy-five years.

References

Barcia, Gonzalez de. *Ensayo cronologico para la historia general de la Florida.* Madrid, 1723.

Briggs, Henry. "Treatise of the North-West Passage to the South Sea, through the Continent of Virginia, and by Fretum Hudson." In Samuel Purchas, *Purchas His Pilgrimes.* London, 1625.

"Codex Magliabechiano." National Library of Florence, Miscellanea XIII, 89 (3). English Translation by Joseph G. Cogswell in the *Collections of the New-York Historical Society*, 1841.

"Copia di una lr̃a. di Giovanni da Verrazzano al Chr^{mo} Re Fran^{co} Re di Francia, della terra plui scopta in nomé di S. M^{ta}." Ms. Ottoboniano 2202, Vatican Library.

Coronelli, Vincenzo. *Atlante Veneto.* Venice, 1690.

Cortés, Hernán. *Praeclara Ferdina[n]di Cortesii de noua maris oceani Hyspania narratio.* Nuremberg and other cities, 1524.

Enciso, Martin Fernandez de. *Suma de geographia.* Seville, 1519.

Hakluyt, Richard. *Discourse Concerning Western Planting.* 1584.

———. *Divers Voyages Touching the Discovery of America.* London, 1582.

Lederer, John. *The Discoveries of John Lederer.* Translated by William Talbot. London, 1672.

Martyr, Peter. *De orbe nouo.* 1516.

Münster, Sebastian. *Cosmographia universale.* Basel, 1544.

Ptolemy. *Geographia.* Rome, 1507–8.

———. *Geographia.* Basel, 1540.

Reisch, Gregor. "Typus universalis terre iuxta modernorum distinctionem et extensionem per regna et provincias." In his *Margarita philosophica nova.* Strasbourg, 1515.

Smith, John. *Generall Historie of Virginia, New-England, and the Summer Isles.* Several editions. London 1624–32.

———. *A True Relation of Such Occurrences and Accidents of Noate as Hath Hapned in Virginia.* London, 1608.

Verrazzano, Giovanni de. *Cèllere Codex*. Pierpont Morgan Library, New York, Morgan Ms.MA.776.

Vespucci, Amerigo. *Paesi nouamente retrouati et Novo Mondo.* Edited by Fracanzano da Montalboddo. Vicenza, 1507.

SELECTED READINGS

Cumming, William P. *The Southeast in Early Maps. . . .* Princeton, N.J.: Princeton University Press, 1958.

Cumming, William P., ed. *The Discoveries of John Lederer.* Charlottesville, Va.: University of Virginia Press, 1958.

Harrisse, Henry. *The Discovery of North Americai.* Amsterdam: N. Israel, 1961.

Murphy, Henry C. *Voyage of Verrazzano: A Chapter in the Early History of Maritime Discovery in America.* New York, Albany: J. Munsell, 1875.

Stephenson, Richard W., and Marianne M. McKee, eds. *Virginia in Maps.* Richmond, Va.: The Library of Virginia, 2000.

Wroth, Lawrence C. *The Voyages of Giovanni da Verrazzano, 1524–1528.* New Haven, Conn.: Yale University Press, 1970.

Fig. 9. Giovanni di Pier Andrea Di Bernardo da Verrazzano (?1485-?1528). Copperplate engraving by F. Allegrini, 1767, from a drawing by Giuseppe Zocchi. Photograph courtesy of The Pierpont Morgan Library New York, MA 776.

The line of descent of Giovanni da Verrazzano indicated on the engraving is incorrect.

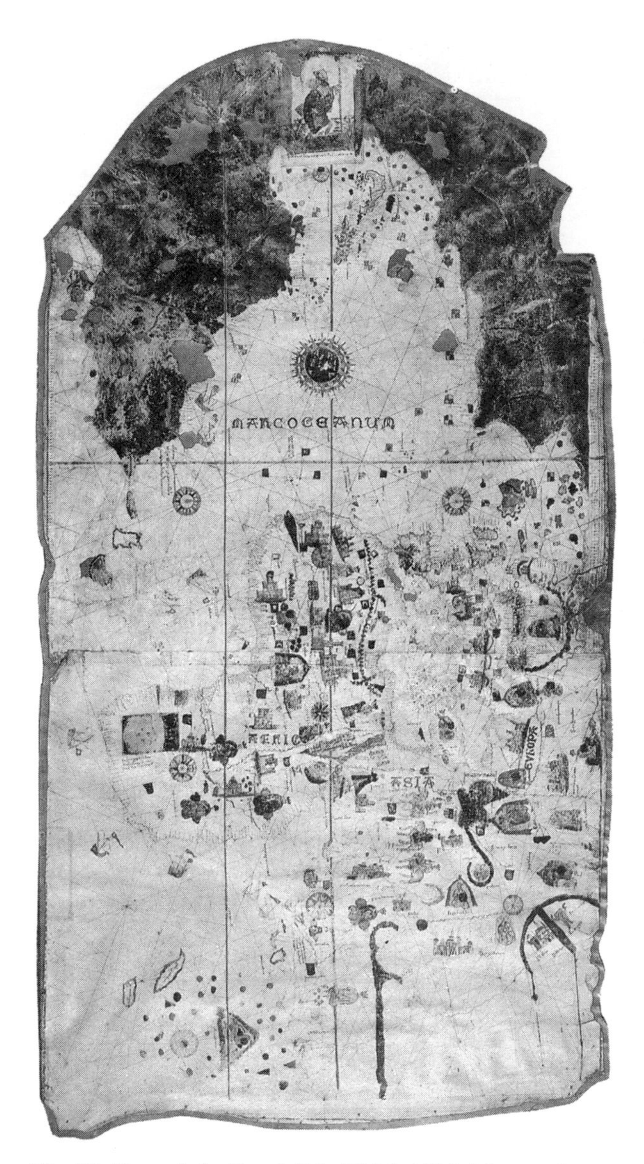

Fig. 10. [Juan de la Cosa]. [No Title]. 1500. Manuscript on oxhide, pen and ink and watercolor, 1800 x 960mm. Photograph from Museo Naval, Madrid, Spain; reproduced with permission.

The earliest extant map showing any part of the North America continent. The phrases "mar descubierta por yngleses" (seas discovered by the English) and "cavo de ynglaterra" (cape of England) indicate evidence of Cabot's voyages of 1497 and 1498.

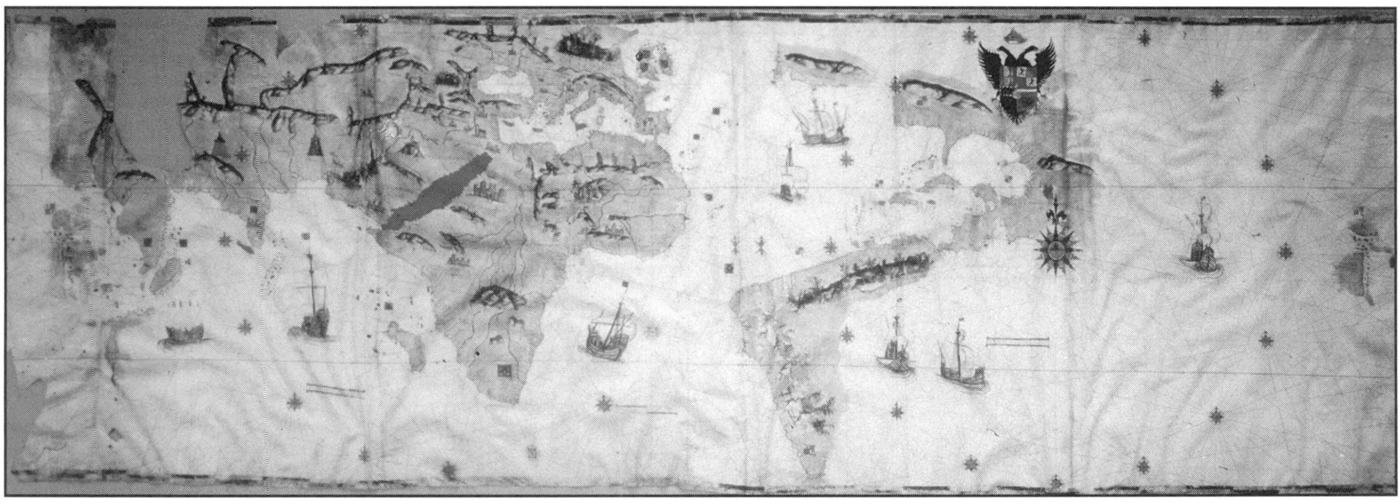

Fig. 11. Juan Vespucci. [No Title]. World chart, manuscript, dated and signed "Ju° Vespuchi piloto de sus mata. me fazia en Seujll año d 1526." Photograph from Hispanic Society of America, New York; reproduced with permission.

The map was drawn by Juan Vespucci, nephew of Amerigo, who, according to Peter Martyr, made several voyages to the American coast. On the coast of current North Carolina, a river is named "R. da Sa Terazanas."

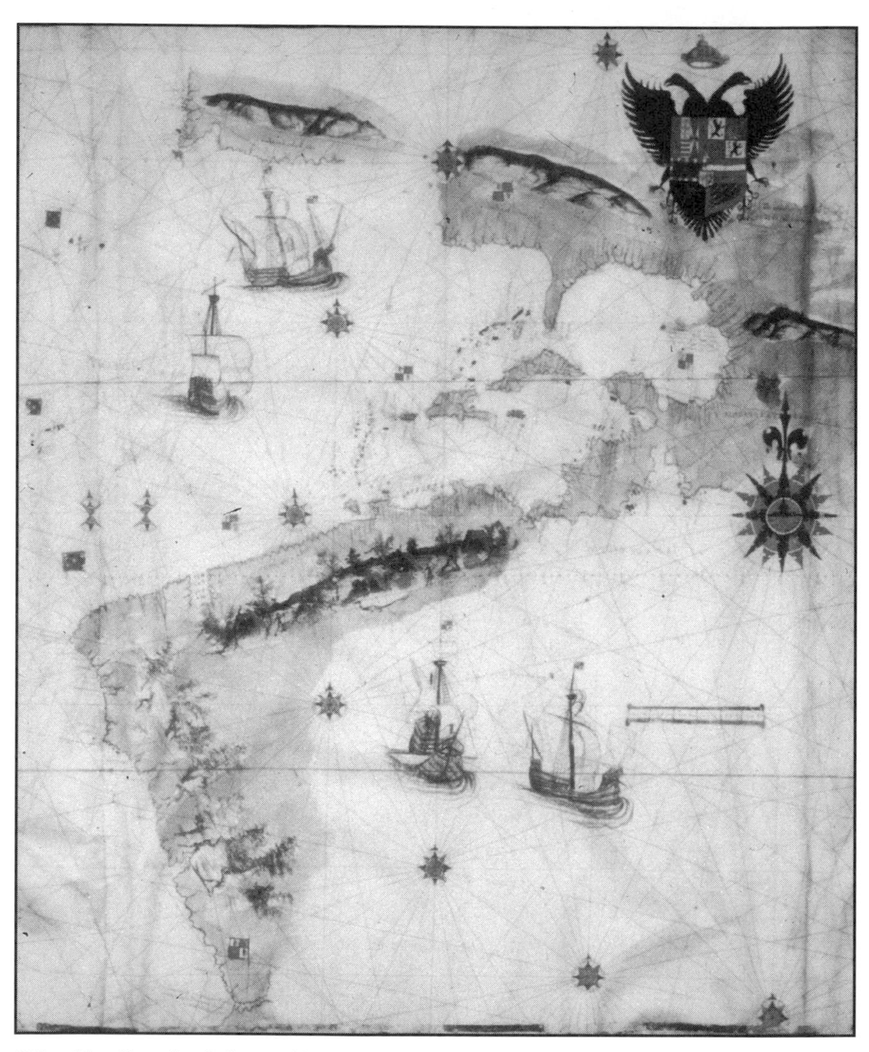

Fig. 11a. Detail of figure 11.

Fig. 12. Vesconte de Maggiolo. [No Title]. World chart, manuscript, 1527. Detail. Photograph courtesy of John Carter Brown Library, Brown University, Providence, R.I.

This is a photograph of the original manuscript map, which was destroyed by the bombing of the Biblioteca Ambrosiana in Milan in World War II. Drawn three years after Verrazzano's voyage, it depicts a narrow strip of land on the Carolina coast, bordered on the west by the Mare Indicum (false Sea of Verrazzano).

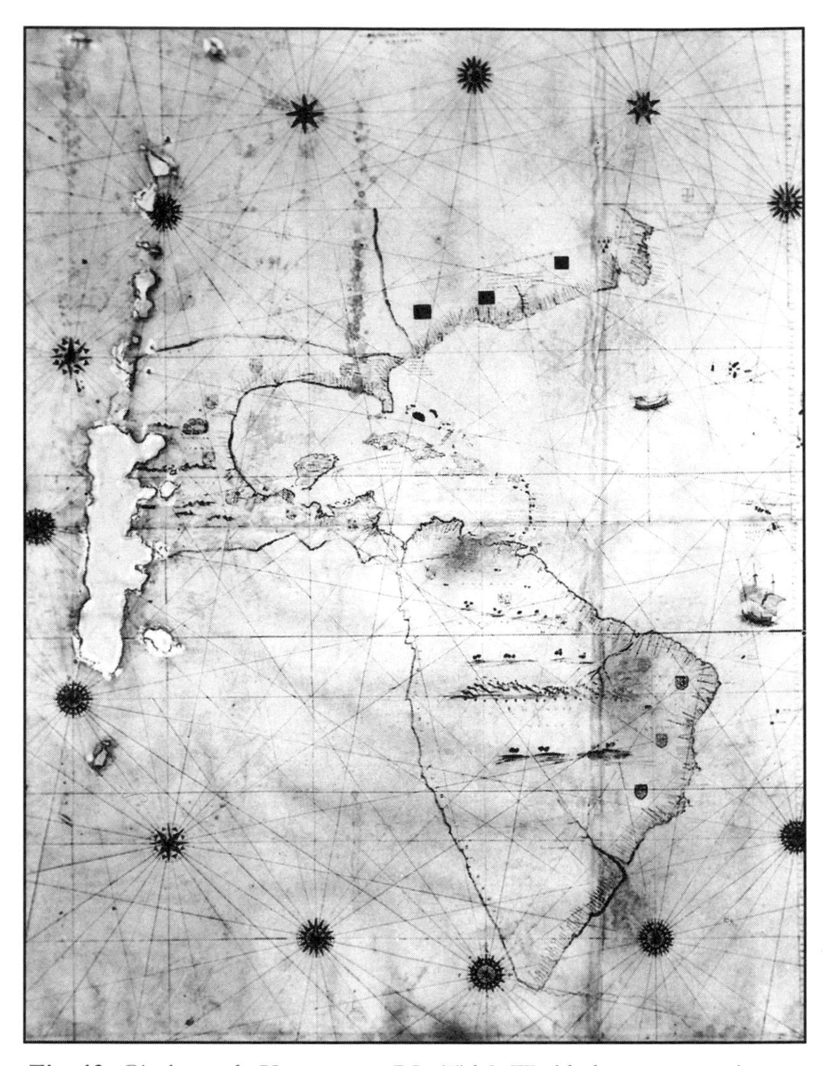

Fig. 13. Girolamo da Verrazzano. [No Title]. World chart, manuscript on parchment, 1529. Detail, entire dimensions 1300 x 2600 mm. Photograph courtesy of Biblioteca Apostolica Vaticana.

The date is based on an inscription over three French flags on the Atlantic coast of North America indicating that the voyage of Verrazzano took place five years prior to the map's execution. The map depicts the false Sea of Verrazzano.

Fig. 14. Robert de Bailly. Copper Globe, 1530. 140 mm. diameter. Photograph courtesy of The Pierpont Morgan Library, New York.

Shows the false Sea of Verrazzano. Follows nomenclature of Girolamo da Verrazzano's map drawn a year earlier. North America is called "Verrazana."

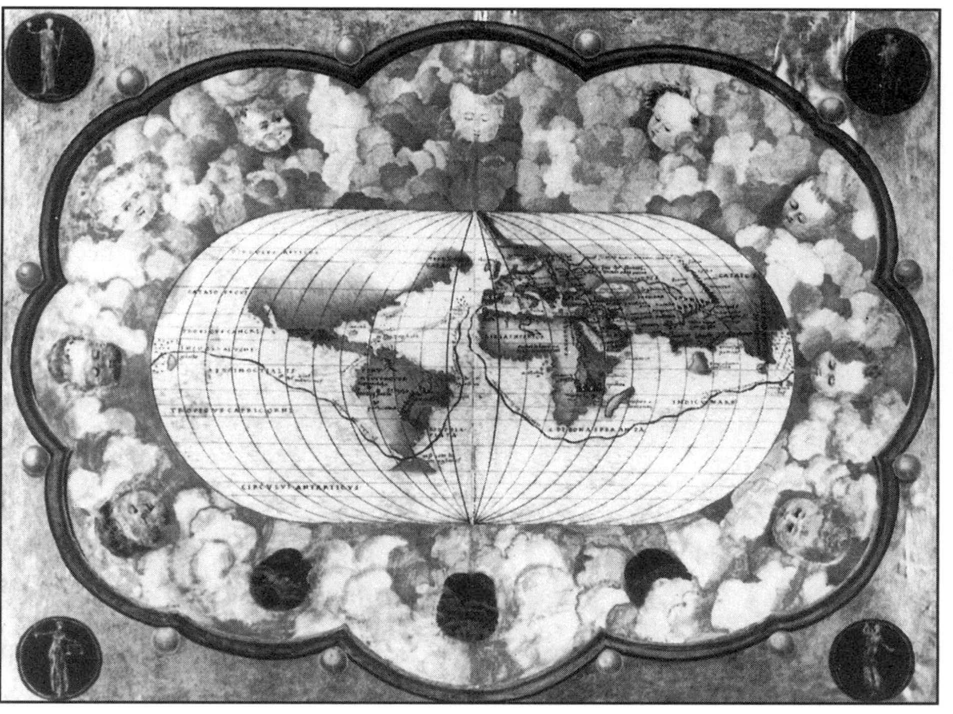

Fig. 15. Battista Agnese. [No Title]. World map, manuscript on vellum, pen and ink and watercolor, 1543–44. 220 x 280 mm. Photograph courtesy of John Carter Brown Library, Brown University, Providence, R.I.

The map shows the false Sea of Verrazzano. It also depicts a distinct California peninsula and bay but leaves the west coast of North America without a defined coast line.

Fig. 16. Sebastian Münster. "NOVAE INSVLAE, XVII NOVA TABVLA." Woodcut, 240 x 340 mm. From Ptolemy, *Geographia* (Basel, 1540). Private Collection.

First map depicting the New World as a distinct insular landmass. It stresses the continuity of the northern and southern continents. The false Sea of Verrazano is apparent.

Fig. 17. Georgio Calapoda. "Florentine Goldsmith's Map." Copperplate, 290 x 390 mm. Derived from an original manuscript on vellum in a portolan atlas of Calapoda's. Photograph courtesy of John Carter Brown Library, Brown University, Providence, R.I.

The map shows a false Sea of Verrazano and is also the first printed map to depict a Lower California peninsula.

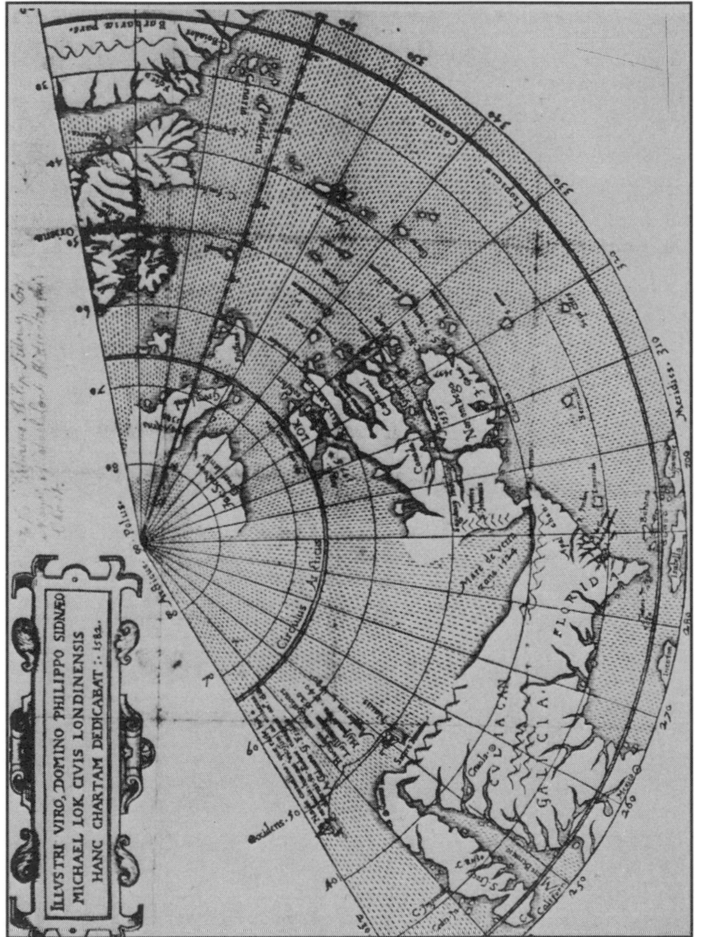

Fig. 18. Michael Lok. [No Title]. Dedication: "Illvstri viro, Domino Philippo Sidnæo Michael Lok civis Londinensis hanc chartam dedicabat: 1582." Woodcut, 290 x 380 mm. From Richard Hakluyt, *Divers Voyages Touching the Discoverie of America* (London, 1583). Private Collection.

Shows "Mare de Verrazana 1524." The map was prepared to promote exploration of a Northwest Passage.

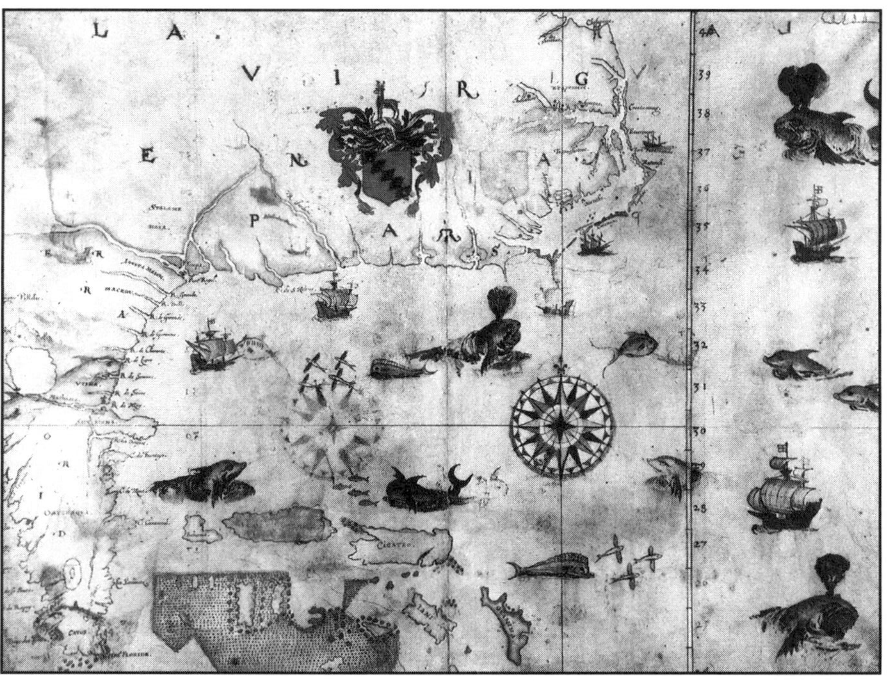

Fig. 19. John White. [No Title]. "La Virgenia Pars." 1585. Manuscript, pen and ink and watercolor, 370 x 470 mm. Photograph courtesy of Department of Manuscripts, British Library, London.

The map shows the southeastern coast from Chesapeake Bay to the tip of Florida. A channel runs from Port Royal westward to a large body of water (Sea of Verrazzano), suggesting a passage to the Pacific Ocean.

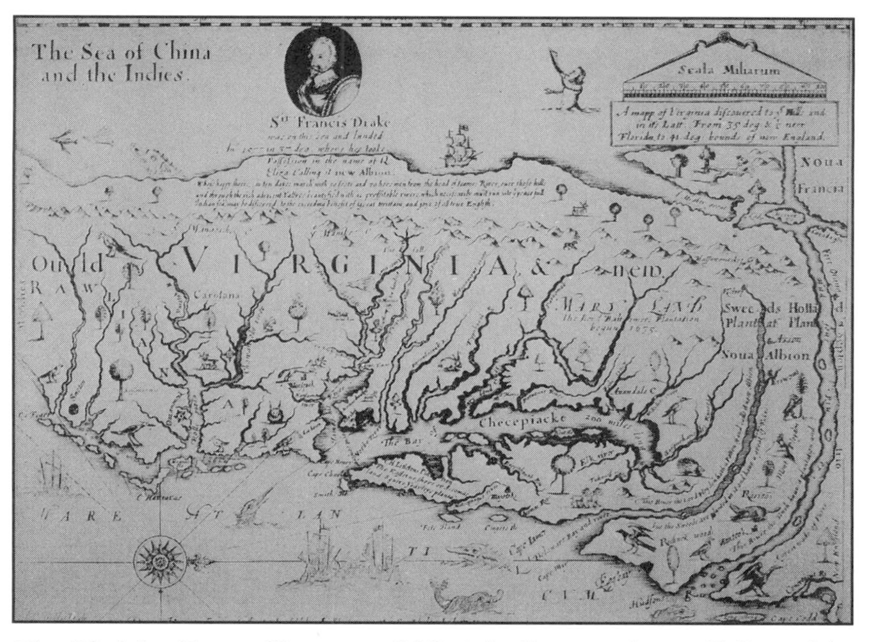

Fig. 20. John Farrer. "A mapp of Virginia discouered to y[e] Falls, and in it's Latt: From 35° deg: & ½ neer Florida to 41 deg. bounds of new England." Copperplate, 270 x 355 mm. From Edward Williams, *Virgo Triumphans, or Virginia Richly and Truly Valued* (London, 1650). Private Collection.

A legend on the map indicates that it is a ten-day march across land from the Atlantic Ocean to the Sea of China. Also, the Hudson River empties into a "Mighty great lake" that is separated by a narrow land bridge from the Sea of China.

CHAPTER THREE

Depicting a Desire

In the same fashion that the Verrazzanian "False Sea" and the associated narrow isthmus in the middle of the North American continent provided an attractive geographic scenario, the Northwest Passage depicted a desirable situation for commerce between Europe and the Far East. Following the Spanish and French incursions in the northern portion of the New World during the second half of the sixteenth century, England and Holland, two nations that were becoming increasingly powerful, also looked to the seas for expansion and acquisition of wealth. Given the absence of a waterway through the North American continent, if one looked westward from London, Southampton, Plymouth, and the ports of Holland, the shortest route to Asia would be along the northern coast of the continental mass interposed between the Atlantic and Pacific Oceans—a Northwest Passage.

A time line of the cartographic history of the Northwest Passage would begin with an expression of a desired direct passage, a desire so strong that, when a distinct waterway was not defined, maps depicted an absence of definition of the northern boundary of the continent, thereby begging the issue. The Northwest Passage was next appropriately deleted from maps based on the contemporarily available facts, and did not regain its rightful position on charts until the twentieth century.

Surely, there is a Northwest Passage that joins the Atlantic and Pacific Oceans. The waters north of North America follow a convoluted course from the Atlantic Ocean to Beaufort Sound, through Coronation Gulf, Queen Maude Gulf, Franklin Strait, Peel Sound, Barrow Strait, Lancaster Strait into Baffin Bay, and then through Bering Strait into the Pacific Ocean (see figure 21).

The history of the mapping of a Northwest Passage is equally convoluted. It began in 1507 with the two Waldseemüller world maps that were included in *Cosmographiae Introductio*. For the first time, these maps depicted a landmass between the Atlantic and Pacific Oceans. Earlier maps presented Newfoundland and Labrador as an extension of Asia. Based on pure speculation, Waldseemüller, on his large main map (see figure 4) and on the small inset map (see figure 4a) of the Western Hemisphere, included a body of water north of the North American continent joining two oceans. The same speculation pertained to the gore map that was structured to fit a small sphere (see figure 5). The 1512 map of Jan ze Stobnicy, a frank plagiarism of the Waldseemüller map, amplified the expanse of water to the north of North America, and positioned Zipangu insula (Japan) relatively close to the west coast of the continent.

In 1527, the British merchant Robert Thorne, in a letter to Henry VIII of England, proposed three more direct routes to the Orient; one eastward along the north coast of Europe, another westward north of the North American continent, and the third across the Polar Sea. A woodcut world map included in Richard Hakluyt's *Divers Voyages* of 1582 had been drawn by Robert Thorne in 1527, according to the text beneath it. On "A Map of the World in Two Hemispheres" of 1538 (see figure 8) by Gerard Mercator, a vague suggestion of a Northwest Passage appears. Sebastian Münster's 1540 woodcut "Novae Insulae, XVII Nova Tabvla (see figure 16), the first map to depict clearly the New World as an insular mass, shows a northern coastline in North America, and Japan (Zipangi) close to the west coast at the latitude of "Temistitan" (Mexico City).

The 1544 map by Gemma Frisius and Peter Apian, published in Antwerp, presents an elongated passage north of "Baccalearum" (North America). By contrast, the map that depicts the Western Hemisphere in Giovanni Battista Ramusio's *Terzo Volume delle Navigationi et Viaggi* in 1556 shows an ill-defined northern part of North America. The same pertains to the Diégo Gutierrez–Hieronymus Cock first wall map of the New World, published in 1562.

One of the more enticing maps for those dedicated to a northern route to the Orient was published in Venice in 1566 by Paolo Forlani. Three important elements appear on a map for the first time (see figure 22). First, there is a narrow body of water between Greenland and North America. Second, a "Mare Setentrionale In Cognito" is present north of the continent. Third, the "Mare Setentrionale" connects via a "Streto de Anian," which took its name from the writings of Marco

Polo, to the Gulf of China. By contrast, Abraham Ortelius's "Americae sive Novi Orbis, Nova Descriptio" (see figure 23) from the first "modern" atlas, published in 1570, has the northern part of North America abutting the map's border, thereby avoiding the issue of a Northwest waterway.

In 1576, a crude map (see figure 24) by Sir Humphrey Gilbert, the Oxford educated privateer and half-brother of Walter Raleigh, indicated that by sailing due west from Greenland and then through a Northwest Passage, one would reach the spice-rich isles of Moluccas. Gilbert based his cartographic declaration on a conversation that he heard in 1568 during which Salvaterra told Sir Henry Sidney that, more than eight years previously, one Andrew Urdaneta had told him (Salvaterra) that he came from the South Sea into Germany through the Northwest Passage. In 1583, Gilbert led an expedition to Newfoundland, which he claimed for England. Newfoundland regards August 5, 1583, as the birth date of the British Commonwealth in the New World. The fortnight spent by members of Gilbert's expedition at St. John's marked the first temporary English settlement in America. Sadly, Gilbert was lost at sea on the return voyage to England.

It was against the backdrop of this cartographic confusion that the first search for a Northwest Passage was conducted. The initial organized quest for the passage took place in 1576 under the command of Martin Frobisher, the dour captain of an English privateer, who had been involved in many questionably illegal activities. As reported in Jeannette Mirsky's *To the Arctic! The Story of Northern Exploration from Earliest Times to the Present,* Frobisher regarded the discovery of the Northwest Passage as "the only thing of the world that was left undone." Several powerful advocates of the Northwest Passage, with the endorsement of Lord Treasurer Burleigh, persuaded the Muscovy Company to relinquish its right to northwestern exploration to a group of merchants who went on to form the Company of Cathay after the completion of Frobisher's first voyage. The importance of the quest was manifest by Queen Elizabeth I, who had subscribed £100 personally, waving from her window in the palace of Placentia at Greenwich as the expedition's three small vessels sailed down the Thames early in the summer of 1576, at the start of their first transoceanic voyage.

One vessel with her crew of four was lost at sea while another returned home after reaching Greenland, leaving Frobisher aboard the *Gabriel* with a crew of eighteen to cross what is now called Davis Strait between Greenland and Baffin Island, which Leif Erikson called Helluland. George Best, in 1578, recorded Frobisher's voyage. Sailing north, Frobisher entered "a great gutte, bay, or passage, deviding as it were two mayne lands or continents asunder." Perceiving the waterway to be a passage to the Orient, Frobisher continued for about 150 miles and, once he had persuaded himself that the body of water did indeed separate Asia and America, he named it "Frobishers Streyts lyke as Magellanus at the Southwest ende of the Worlde, havying discovered the passage to the South Sea . . . called the same straites Magellanes streights."

In this area, now appropriately called Frobisher Bay, the explorers encountered Inuit, and five of the crew were lost on shore. One of the remaining crew brought back from what is now Hall's Inlet a piece of black rock that would serve as the stimulus for additional explorations. The ship returned to England in October with the rock, a Native, and a kayak.

Anticipation that the black rock represented a source of gold led to the rapid outfitting of a second expedition by Frobisher under the auspices of the newly organized Company of Cathay, to which Queen Elizabeth subscribed £1,000 and furnished a tall ship that carried a compliment of between 115 and 120 men, mainly miners. The prime goal was to mine the ore, but also to rescue the five sailors who had been taken prisoner during the first voyage. Specific instructions were issued to emphasize that the search for a Northwest Passage was to be conducted only if it did not interfere with the other two priorities. The fleet of three vessels made landfall at Hall's Island in Frobisher Bay on July 15, 1577. The summer was spent mining about two hundred tons of ore and loading it on the ships. Some of the mining took place on what is now called Kodlunarn's (White Men's) Island. The ships returned to England in mid-September, and the ore was stored awaiting completion of the building of furnaces at Dartford along the lower Thames to refine it. No gold would ever be extracted; the rock was truly "fool's gold."

Before the status of the ore was defined, Frobisher left England in the early summer of 1578 with a fleet of fifteen vessels on his third expedition to the area to continue mining the ore. The ships sailed to "Frobisher Straightes" and several of the vessels entered what we now call Hudson Strait, which Frobisher named "Mistaken Straightes." After they were beset with storms, the fleet reassembled, and, in the region of Countess of Warwick Sound, mined and loaded over 1000 tons of the black ore onto the ships before returning home.

The first account of the three voyages of Martin Frobisher appeared in London in 1578 in a book entitled *A True Discourse of the Late Voyages of Discoverie for the Finding of a Passage to Cathaya.* The book contains a woodcut map (see figure 25) by the author, George Best, and the probable map maker, James Beare, both of whom sailed on all three expeditions. A separate chart in the book was probably drawn by James Beare. It is an expanded cartographic representation that depicts "Frobisher Streightes," the "Mistaken Straightes," and a channel bearing the words "The Way Trendin to Cathia" (see figure 26).

Because of the failure to extract gold from the ore, Frobisher's reputation declined, but only temporarily. In 1585, he served as Vice-Admiral under Sir Francis Drake on an expedition against the Spanish West Indies. Frobisher covered himself with glory in the battle with the Spanish Armada during which he commanded the *Triumph*, the Royal Navy's biggest ship. He was rewarded with a knighthood.

Michael Lok, the Secretary of the Muscovy Company of London, which had been chartered in 1555 to reach Cathay by a Northeast Passage, north of Russia, was a major financier of the Frobisher voyages. As a consequence, Lok was bankrupted and spent time in a debtors prison. He had acquired a copy of the map by Girolamo da Verrazzano that depicted the findings of the cartographer's brother, Giovanni. Lok produced a map (see figure 16) that appeared in Richard Hakluyt's first publication, the *Divers Voyages* of 1582. The map, which incorporated the "False Sea" and narrow isthmus in the middle of North America, was used to promote the theory of a possible Northwest Passage. The map presents evidence of the Frobisher voyages, and "Angli 1576" appears on an island where the ore was mined.

In 1595, in Venice, Michael Lok met Apostolos Valerianos, a Greek sea captain who, while sailing for Spain, was known as Juan de la Fuca. Juan de la Fuca narrated that, while cruising along the west coast of North America in 1592, he came upon a strait at 47° North Latitude and that he sailed easterly through that body of water to the Atlantic Ocean. The protagonist of this apocryphal tale has his name, Juan de la Fuca, imprinted on modern maps, designating the strait between the northwest coast of the state of Washington and Vancouver Island, linking Puget Sound and the Strait of Georgia with the Pacific Ocean.

Richard Hakluyt's *Divers Voyages*, published in 1582, included Lok's map, which suggested the presence of a Northwest Passage, and Hakluyt himself wrote of the "great probabilitie, of passage by the North-west part of America." Continued enthusiasm for the discovery of a northern passage to Cathay led to the formation of a new syndicate headed by William Sanderson, which sponsored the three voyages conducted by John Davis in 1585, 1586, and 1587. These expeditions, conducted by an accomplished and scientifically inclined seaman, for whom the strait between the southwest coast of Greenland and Baffin Island leading into Baffin Bay would be named, set the stage for the later discoveries of Henry Hudson and William Baffin.

On his first voyage, Davis sailed with two vessels from Greenland across Davis Strait to the Cumberland Peninsula on Baffin Island, where the easternmost point was named "Cape Walsingham" for the sponsor, Sir Francis Walsingham, Secretary of State of England. On that voyage, chronicled with the other two in his 1595 *The Worldes Hydrographicall Discription*, Davis named a body of water "Exeter Sound" after the city that sponsored the mission, and another body of water "Clifford Sound" after the Earl of Cumberland. The ships sailed for a distance of about 180 miles into Cumberland Sound before returning home, where Davis reported "that the north-west passage is a matter nothing doubtfull, but at any tyme almost to be passed."

In 1586 Davis returned with four ships to the same area. On his return to England, he declared that the passage "must be in one of foure places or els not at all." These four possibilities were Cumberland Sound, Hudson Strait, Hamilton Inlet, or north up Davis Strait, explored by Davis. On the third voyage in 1587, several men were honored by having their names attached to various locations. A headland was called "Hope Sanderson" after Davis's patron, "Lumley's Inlet" took the name of the high steward of the University of Oxford, and "Cape Chidley" honored a Devon mariner. On his return from the third voyage, Davis wrote in *The Worldes Hydrographicall Discription*: "I have bene in 73 degrees, finding the sees all open. . . . The passage is most probable, the execution easie, as at my comming you shall fully know." Davis drew from his experience to endorse the theory that there was "a passage by the Norwest, without any land impediments to hinder the same."

During the final decades of the sixteenth century, the findings of Frobisher and Davis and the concept of a Northwest Passage found expression on several maps. Rumold Mercator's double hemispheric world map, a condensation of his father's (Gerard) great world map of 1569 (see figure 27), was published in 1587. It depicts a Northwest Passage, as do Abraham Ortelius's "Typus Orbus Terrarum," published the same year, and the 1588 map in Sebastian Münster's *Cosmographia*. Nowhere does a Northwest Passage stand out more boldly than in *Speculum Orbis Terrae. . .* by Cornelis de Jode, published in 1593 (see figure 28). In that atlas, the passage is apparent on the two new world maps and on the map devoted to North America, which incorporates a direct Northwest Passage running the length of the top of the map with a "Lago de Conibas" emptying into it.

The same geographic representation appears on Michael Mercator's 1595 "America sive India Nova" and Theodore de Bry's "America sive Novus Orbis . . ." published in 1596. In 1597, Cornelis van Wytfliet's *Desciptiones Ptolemaicae Augmenum* was published, representing the first atlas that focused on all of the Americas. Included is a map entitled "Conibas Regio cum Vicinis Gentibut," showing a "Lago de Conibas" emptying into a Northwest Passage (see figure 29). The native village of Hochelaga on the site to become Montreal also appears on that map. Another map in the same atlas concentrates on the Labrador region (see figure 30). Frobisher's Bay is shown as a strait transecting the southern tip of Greenland while the turbulent waters characterizing the current-day Hudson Strait are suggested by the label "a furious over fall."

At the end of the sixteenth century and the beginning of the seventeenth century, ambivalence about the presence of a Northwest Passage was manifest on maps. The famous Wright-Molyneaux map of 1599, present in some copies of Richard Hakluyt's *Voyages, Navigations, and Discoueries of the English Nation* . . . begs the issue by leaving the region north of the continent blank (see figure 31). A similar representation appears on William J. Blaeu's 1606 world map, Jodocus Hondius's world map of 1608 and Pieter van den Keerre's world map of 1608. By contrast, Arnold Arnoldi's two-sheet world map of 1601 shows a direct passage from the Atlantic to the Pacific Ocean, as does Jean le Clerc's map engraved by Jodocus Hondius in 1602 and Petrus Plancius's world map of about 1605. A Northwest Passage is conspicuously absent from all major world maps for the remainder of the first half of the seventeenth century.

Among the early seventeenth-century maps focusing on North America, a similar circumstance regarding the ambivalent mapping of a Northwest Passage is manifest. Gabriel Tatton, a noted London hydrographer, published two elegant maps in 1600. Both "Maris Pacifici" (see figure 42) and "Noua et Rece Terrarum et Regnorum Californiae nouae Hispaniae Mexicanae, et Peruviae . . ." cut the North American continent off at the top of the printed page in order to suggest that the issue of a passage was not defined. Although a sea connecting the Atlantic and Pacific Oceans appears on the map of the North Pole published by Jodocus Hondius in an atlas in 1607, a complete Northwest Passage is absent on all his other major maps depicting only North America. The title page of Michael Colijn's 1622 edition of *Descriptio Indiae Occidentalis* . . . by Antonio de Herrera y Tordesillas contains a small map of the Western Hemisphere. The map (see figure 43, which shows the map on the title page of the 1623 German edition) depicts a Northwest Passage, but its chief impact was as the first map to present California as an island. Thus, with occasional exceptions, early in the seventeenth century, the Northwest Passage was *deleted* from maps.

During the early part of the seventeenth century, three English explorers, Henry Hudson, Thomas Button, and William Baffin, continued the effort to find the Northwest Passage. The Muscovy Company hired Henry Hudson to reach Cathay by way of a Northeast Passage via the North Pole. Following this unsuccessful venture, he was recruited by the Dutch East India Company under whose sponsorship, in September 1609, he sailed the *Half Moon* up the river that bears his name to a point just below the present site of Albany, New York. The voyage, which actually began as an attempt to find a *Northeast* passage to Cathay and was thwarted by the ice, eventuated in the Dutch establishing a foothold in North America in 1614. In 1626, Peter Minuit purchased the island of Manhattan from the Natives for the equivalent of twenty-four dollars in goods.

Henry Hudson became interested in the Northwest Passage as a consequence of reading Captain George Waymouth's log books of an unsuccessful probing of the waterway north of North America in 1602 under the auspices of the newly founded East India Company. Hudson gained the sponsorship of a group of English courtiers to map the Northwest Passage. He sailed aboard Waymouth's old ship, the *Discovery,* which was to be credited with a total of six Arctic voyages. In 1610, Hudson crossed the Atlantic Ocean a second time, sailed through Davis Strait, and entered the strait and bay that both now bear his name. He proceeded to follow the eastern shore of Hudson Bay, which he believed to be the Pacific Ocean, and wintered in James Bay. During that voyage, Hudson named two islands, Nottingham and Digges, and Cape Wolstenholme for his backers. In June 1611, mutineers took over the vessel and set Hudson, his son, and six crew members adrift in an open boat. No member of that group was ever heard of again. The ship returned to England under the command of Robert Bylot.

Hessel Gerritsz engraved a chart of Henry Hudson's discoveries (see figure 32) and published it in *Descriptio ac delineatio geographica detenctionis freti*, Amsterdam 1612. The map depicts Hudson Strait, the opening into Hudson Bay, and James Bay. The names in Hudson Strait are those bestowed by Hudson, while the "Mare Magnum" (wide sea) at the western extreme of the chart illustrates Hudson's conviction that he had completed a Northwest Passage to the South Sea.

In 1612, the *Discovery* returned to Hudson Bay under the command of Sir Thomas Button with the sponsorship of the newly chartered Company of the Merchants of London Discoverers of the North-West Passage. Included among the patrons were the Archbishop of Canterbury, the Solicitor General Sir Francis Bacon, the historian Richard Hakluyt, and the mathematician Henry Briggs, whose map would make California an island on maps for over a hundred years. Button, with Robert Bylot as his navigator, entered Hudson Bay and sailed across, for the first time, to the western shore, where, as evidence of a passage, he noted the tides. The ship anchored at the mouth of a river that he named the *Nelson* after a member of the crew who died there. This was the first time that Englishmen wintered in the region of what would become York Factory, the major site of the Hudson Bay Colony. On his return to England, Button declared that he believed there was a passage with the major flow between current Salisbury Island and Nottingham Island into Foxe Channel.

After William Baffin and Robert Bylot again sailed through Hudson Strait and found no passage to the north, Baffin continued to reaffirm the presence of a Northwest Passage with the statement, transcribed in Clements R. Markham's *The Voyages of William Baffin, 1612–1622*:

> Doubtles theare is a passadge. But within this strayte whome is called Hudson's Straytes, I am doubtfull, supposing the contrarye . . . we haue not beene in any tyde then that from Resolutyon Iland [named for Button's ship], and the greatest indraft of that commeth from Dauis Strayts, and my judgement is . . . the mayne [passage] will be upp fretum Dauis.

The following year, the same two veteran Arctic explorers returned to the region with the explicit instructions from the Company to sail north to 80° and "then direct your course to fall in with the land of Yedzo [Japan]." They sailed north to Baffin Island and up Baffin Bay, both appropriately named for William Baffin, who made a total of six Arctic voyages. At 76° North Latitude, they entered Whale Sound, which they named for the great number of whales, and "Sir Thomas Smith's Sound" [Smith Sound]. They then crossed the openings of "Alderman Jones's Sound" and "Sir James Lancaster's Sound," the very area that would be established as a Northwest Passage in the nineteenth and twentieth centuries.

The only exploration for a Northwest Passage during the seventeenth century that was not an English effort was conducted under orders from King Christian IV of Denmark. Jens Munk, who served in the Royal Danish Navy and who had previously tried to locate a Northeast Passage, was charged with plotting the Mercator projections of a Northwest Passage westward from Hudson Bay. Munk departed Denmark with two naval ships on May 30, 1619, and the party landed at the mouth of the Churchill River, where they wintered. Scurvy was so rampant among the crew that by summer only Munk and two others remained alive. They refloated one of the vessels and, incredibly, succeeded in sailing the ship safely back to Copenhagen.

Munk detailed his exploration and chilling tale of suffering and endurance in *Navigatio septentrionalis*, Copenhagen 1624. The book includes one of the most vivid descriptions of scurvy, a disease that would plague mariners until Dr. James Lind defined the cause and treatment in 1754 in the first prospective randomized controlled clinical trial in the history of Medicine. Munk wrote:

> As regards the symptoms and peculiarities of the illness which had fallen upon us: it was a rare and extraordinary one. Because all the limbs and joints were so miserably drawn together, with great pains in the loins, as if a thousand knives were thrust through them. The body at the same time was blue and brown, as when one gets a black eye, and the whole body was quite powerless. The mouth was also is a very bad and miserable condition, as all the teeth were loose, so that we could not eat any victuals . . . .

Information gained during explorations in the early seventeenth century found expression on maps. Samuel de Champlain's 1616 map (see figure 52), which was embellished by Pierre Duval in 1654, depicts a speculative passage to "Japan and China." The 1625 map (see figure 45) of mathematician Henry Briggs, which would erroneously redefine the cartography of California by making it an island, was actually created to illustrate his "Treatise of the North-West Passage to the South Sea, through the Continent of Virginia, and by Fretum Hudson," published in *Purchas His Pilgrimes*, London 1625. Briggs's stated hypothesis was that Hudson Strait and Button Bay provided "a faire entrance to ye nearest and most temperate passage to Japa & China." Briggs's was the first to place the name "Hudsons bay" on a printed map, although it is incorrectly applied to James Bay, while "Buttons Baie" is assigned to the western part of Hudson Bay.

Six years later, Luke Foxe set out to check Brigg's hypothesis. Foxe, with customary dash, as evidenced by his memoirs published in 1635 and entitled *North-West Foxe*, left London in the spring of 1631 and sailed around Hudson Bay, where he reported a flood tide of over eighteen feet at a latitude of 64° 10' north, suggesting the inevitability of a Northwest Passage. He did, however, conclude that there was no westward passage out of Hudson Bay itself and that, if such a passage were present, it must be by "Sir T. Roes Welcom" in 65° North. Foxe's name would remain on maps as Foxe Basin.

Paralleling and intertwining with Luke Foxe's voyage was the exploration by Captain Thomas James, whose name would persist on a large bay in the Arctic. James was a well-educated barrister whose vivid descriptions in his memoirs, *The Strange and Dangerous Voyage of Captaine Thomas James,* London 1633, inspired some of Samuel Taylor Coleridge's verbal images in *The Rime of the Ancient Mariner.* The three stanzas referred to are:

> And now there came both mist and snow,
> And it grew wondrous cold:
> And ice, mast-high came floating by,
> As green as emerald.
>
> And through the drifts the snowy clifts,
> Did send a dismal sheen:
> Nor shapes of men nor beasts we ken—
> The ice was all between.
>
> The ice was here, the ice was there,
> The ice was all around:
> It cracked and growled, and roared and howled,
> Like noises in a swound!

Like Foxe, James left London in the spring of 1631 and headed for an identical destination where the two met on August 29 in Hudson Bay. Whereas Foxe returned to England the same year, James wintered in the bay which would be assigned his name. Also like Foxe, James's exploration of the west coast of Hudson Bay led to the conclusion that there was no practical passage from the bay to the South Sea. Subsequent to these two parallel voyages, interest in the maritime exploration of the geographic holy grail, the desired Northwest Passage, faded and lay dormant for many years.

But that which was erased could always be reinstated by the acquisition of more knowledge. In the case of the Northwest Passage, the facts, which resulted in the reinsertion of the passage on maps, did not become conclusive until early in the twentieth century. The three intervening centuries, however, did not constitute a period of inactivity along the northern coast of North America.

Initially, additional information relating to a potential Northwest Passage next came not from the seas but rather from the land, and were based on the commercial interests of the Hudson's Bay Company. The central figure in the genesis of the Company was Prince Rupert, a Renaissance polymath, nephew of Charles I, cousin of Charles II, and a member of the Restoration Court. Stimulated by the commercial potential that arose during personal conversations in London with a pair of Canadian fur traders (*coureurs de bois*), Pierre-Esprit Radisson and Médard Chouart, Sieur Des Groseillers (with the mnemonic designation "Radishes and Gooseberries"), Prince Rupert organized a syndicate named the Company of Adventurers to capture the North American fur trade from the French.

On June 3, 1668, two vessels equipped by the Company, one bearing Radisson and the other Groseillers, left the Thames heading for the New World. The one with Radisson aboard was forced back by a storm, but Groseillers and the remaining ship, the *Nonsuch,* returned to London in June 1669 with furs that were sold on the London market. The commercial success of the venture led to royal approval of a monopoly over the land encompassing the fur trade. On May 2, 1670, Charles II granted Rupert and his associates a charter for the Hudson's Bay Company, with Prince Rupert as its Governor. The Company, which was designated the "true and absolute Lordes and Proprietors" of a vast territory to be known as Rupert's Land consisting of almost forty percent of current Canada. As stated in the charter, the Company received:

> sole trade and commerce of those Seas Streightes Bayes Rivers Lakes Creeks and Sounds in whatsoever latitude they shall bee that lye within the entrance of the Streightes commonly called Hudson's Streightes, together with all the Lands and Territorys upon the Countryes Coasts and confynes of the Seas Bayes Rivers Lakes Creks and Sounds aforesaid that are not actually possessed by or granted to any of our subjectes or possessed by Subjectes of any other Christian Prince or State.

The Hudson's Bay Company thus joined the Swedish mining company, Stora Kopparberg (1288), the Löwenbrau Brewery (1383), the Banco di Napoli (1539), and Joseph Travers & Sons Ltd. in Singapore (1666) as one of the oldest continuing corporations, and it is the only one that has maintained uninterrupted operation under the same name and organizational structure. Three wooden forts were built on James Bay, the southern extension of Hudson Bay, in 1675, and the Hudson's Bay Company developed a most profitable fur trade with the regional Indians.

Although the Hudson's Bay Company did not regard the search for a Northwest Passage as a major focus, the Company did sponsor a mission led by James Knight in 1719. The aim of the mission was the exploration of the northwest coast of Hudson's Bay. Knight and his crew of forty men left London in June 1719, never to be heard from again.

At this point, the chronology of the Northwest Passage dramatically changed its venue from the eastern gateway to the west. In 1724, Peter the Great, Tsar of Russia, sent an expedition lead by Vitus Bering, a Danish seaman in Russian employ, with instructions to sail north from Kamchatka to "determine where it joins with America" or to define the western outlet of the Strait of Anian, which had appeared on maps since 1566. After a four-year, 6000-mile overland journey from St. Petersburg to the Katmchatka Peninsula, the expedition set sail in July 1728 and cruised through the strait that now bears Bering's name. The vessel's crew failed to sight any part of the American continent.

In 1731, Bering received orders for a second mission, which included a series of overland explorations of Siberia, but had as its main thrust reaching the American mainland. After a long delay, Bering set sail in June 1741 from Kamchatka in the *St. Peter,* accompanied by Aleksi Chirikov in command of the *St. Paul.* The two vessels became separated, never to meet again. Chrikov sighted the American shore when he anchored near the island of Sitka before returning home. Bering, on the other hand, sighted Mt. St. Elias in Alaska and landed at Kayak Island. On the return voyage through the Aleutian islands, the ship was wrecked and many of the crew died, including Bering. According to his surviving journal, edited and published by Frank A. Golder, Bering was convinced that he had landed in America.

The French geographer Joseph Nicolas de l'Isle, younger brother of the famous French cartographer Guillaume, was in St. Petersburg and helped plan the second Bering expedition. When he returned to France in 1752, he and Philippe Buache published two maps. The first shows a large "Mer de l'Ouest" indenting the northwestern portion of the American continent, while the second depicts a smaller indentation but includes a waterway—the apocryphal concept of Admiral de Fonte—from the Pacific Ocean leading to Hudson Bay and Baffin Bay (see page 88) (see figure 33a). By contrast, Gerhard Friedrich Müller's "Nouvelle Carte des Découvertes faites par des Vaisseaux Russiens," St. Petersburg, 1758 (see figure 33b), presented the Russian viewpoint. This map asserted that there was a narrow strait separating Asia and America. The Alaska peninsula was extended far to the west as a continuation of the California coast, thereby minimizing the chances of finding a Northwest Passage through Hudson Bay.

Although the Hudson's Bay Company's charter mentioned the quest for a passage to the Pacific and the Orient, it was essentially neglected by the Company, which was dedicated to the fur trade. In fact, the 1741–42 voyage of Captain Christopher Middleton grew out of the conviction of Arthur Dobbs, a member of the Irish House of Commons, who was convinced that Hudson's Bay Company thwarted further exploration for a Northwest Passage in order to protect its commercial interests. Dobbs was able to entice Captain Middleton, one of the company's ship masters, to conduct a voyage of exploration endorsed by King George II.

Dobbs's conclusion that a Northwest Passage existed may have been influenced by an extraordinary communication that appeared in the *Monthly Miscellany or Memoirs for the Curious*, published in London in 1708. The report consisted of a letter from an "Admiral Bartholomew de Fonte, then Admiral of New Spain and Peru," detailing a voyage that he made in 1640. The letter indicated that his ship sailed north along the west coast of North America to the River los Reyes at 53° N. Latitude. Fonte, with two ships, then proceeded up the easily navigable river to Lake Belle, Lake Fonte, and the Strait of Ronquillo. In that region, they met a merchant ship, which had allegedly sailed from Boston, and exchanged items with the Captain. The 1708 publication received no attention until British interest in a Northwest Passage returned in the 1740s. A Jesuit priest later reviewed documents in Spain and proved that de Fonte's account was entirely false. A search of the Spanish Archives failed to uncover any evidence of the voyage or the existence of an Admiral Bartholomew de Fonte.

Middleton commanded an expedition, consisting of two ships, that left England in June 1741 and wintered at The Hudson's Bay Company's Old Factory on the Churchill River, which enters Hudson Bay from the southwest. During the following summer, the ships sailed north and through Roe's Welcome Sound between the west coast of Southampton Island and the Wager River. The ships proceeded into a bay, the upper portion of which abutted the Arctic Circle. At that point, the ice packs thwarted progress north, leading to the assignment of the name "Repulse Bay." Middleton's expedition proved that Welcome Sound was not an entrance into the Northwest Passage (see figure 34). The ships then returned to England. Dobbs initiated a campaign to discredit Middleton, and was convinced that

Middleton deliberately failed to define a Northwest Passage in order to maintain the monopoly of the Hudson's Bay Company.

In 1744, Dobbs produced *An Account of the Countries Adjoining to Hudson's Bay*, which incorporated the alleged discoveries of Admiral de Fonte as evidence of existence of a Northwest Passage. In 1745, Parliament offered a £20,000 award for the discovery of a navigable Northwest Passage. Dobbs organized a second voyage in 1746–47, led by William Moor, who had been Middleton's second-in-command, but they discovered no new waterway. In 1768, Moses Norton, the Factor (Governor) at Hudson's Bay Company's Prince of Wales Fort at Churchill, presented to the London Committee a sketch drawn by two Chipewyans delineating a river to the northwest of the fort running between copper mines and timberland. The Chipewyans' information led to the authorization of an overland expedition to the alleged copper mines and, even more importantly, to determine whether there was a navigable Northwest Passage along the west coast of Hudson Bay. In 1770–72, Samuel Hearne, accompanied by a group of Chipewyans, undertook a momentous trek, which he chronicled in *A Journey from Prince of Wales's Fort in Hudson's Bay to the Northern Ocean*, London 1795. After traveling across the Barrens of the northern tundra, they followed the Coppermine River to its mouth, which ended in a broad expanse of water north of the Arctic Circle. Thus, Hearne became the first European to view the northern coastline on the American continent, and his journey proved that there was *no* Northwest Passage south of the Arctic Circle. On his map, Hearne placed the mouth of the Coppermine River about two hundred miles too far to the north.

History's famous circumnavigator, Captain James Cook, contributed indirectly to an appreciation of the Northwest Passage. Cook had made his reputation initially in the eastern waterways of North America by charting the Saint Lawrence River at the time of the siege of Quebec in 1759 and also charting Newfoundland, where he served as its Governor. Captain Cook's voyages of 1768–71 and of 1772–75 in the Southern Hemisphere, during which he became the first European to sail south of the Antarctic Circle, were followed by third voyage of 1776–80, which had as its object the discovery of the Northwest Passage by way of the Pacific Ocean. Cook was murdered by natives of the Sandwich Islands (now Hawaii) in 1779, but the charts brought back to London by his officers and published in 1784 mapped the northwest coast of America as far north as the Bering Strait, for the first time in an appropriate fashion. No inland passage from the Pacific Ocean was noted.

The first time a Northwest Passage was made actually took place on land; it was completed on 22 July 1793, when Alexander Mackenzie sighted the Pacific Ocean. Mackenzie was in the employ of the North West Company, a Montreal-based rival of the Hudson's Bay Company. In June 1779, accompanied by four French Canadian voyageurs, a Chipewyan named English Chief, some of his followers and wives, and a young German named John Steinbruck, Mackenzie departed Fort Chipewyan on the southern shore of Lake Athabasca in current Alberta. They reached the large river that now bears Mackenzie's name and followed that river northward over 1000 miles to the Arctic Ocean, which they reached on the very day that the French Revolution broke out, 14 July 1789. The expedition proved that there was not a Northwest Passage below 69° 15' North Latitude. On a second trek, in 1792–93, Mackenzie completed his continental traverse when he ascended the Peace River from Lake Athabaska, crossed the Rocky Mountains, and, following the Fraser River, reached the Pacific Ocean at 52° North Latitude, six weeks after Vancouver reached the same region.

The concept of a Northwest Passage between the Atlantic and Pacific Oceans was laid to rest temporarily during the final decade of the eighteenth century with the three voyages led by Captain George Vancouver. From 1792 through 1794, Captain George Vancouver, who had sailed with Captain Cook on his last two voyages, conducted a survey of the northwest coast of America from 39° North Latitude to Cook Sound, 61° North Latitude. During the voyage, Vancouver Island received the Captain's name, while Puget sound was named for Lieutenant Puget, the second-in-command, and Mount Ranier for the Admiral responsible for the mission. Vancouver wrote in his journal that his explorations "set aside every opinion of North-West Passage . . . existing between the Pacific and the interior of the American continent."

Although cartographic representations of a Northwest Passage remained deleted throughout the eighteenth and nineteenth centuries, the hope that such a passage would be discovered persisted and manifested itself throughout the ninetenth century. England's position of supremacy resulted from its dominance of the seas. From the time of defeat of the Spanish Armada in 1588, the use of England's control of the seas for commercial expansion became a dominant theme. Renewal of Arctic exploration occurred during a time when the British Navy needed a cause after it was no longer consumed by war with Napoleonic France. British sea power had played a significant role in the defeat of Napoleon, particularly when Nelson defeated the French and Spanish fleets at the Battle of Trafalgar in 1805.

During the first half of the nineteenth century, a central figure in the exploration of the waterways in the northern part of the North America was Sir John Barrow, for whom Point Barrow on the north coast of Alaska and Barrow Strait in the eastern portion of the Canadian Arctic waterway were named. Barrow served from 1804 to 1845 as Second Secretary of the Admiralty under First Lord of the Admiralty, Lord Melville, whose name appears on Melville Island, Melville Sound, and Melville Peninsula.

In 1818, based on the premise that "if the great polar basin should be free of land, the probability is, that it will also be free of ice," the Admiralty dispatched two ships to the North Pole and two in search of the Northwest Passage. To Barrow, "the discovery of a north-west passage to India and China has always been considered an object peculiarly British." Barrow offered as evidence of "a free communication between the Atlantic and Pacific" the recovery of harpoons struck off Spitsbergen in the Atlantic in whales killed off the Northwest coast.

The first nineteenth-century expedition to the North Pole was conducted under the command of Captain David Buchan with Lieutenant John Franklin as second-in-command. The name of John Franklin, whose recognition began aboard the *Bellerophon* as a signal-midshipman at the Battle of Trafalgar and increased with his part in the Battle of New Orleans in 1815, would dominate the history of exploration of the Arctic and the quest for the Northwest Passage throughout the remainder of the nineteenth century. Buchan's mission in 1818 was to reach the North Pole and proceed to the Bering Sea or, failing that, to enter Baffin's Bay, probe the northern region, and return to England via Davis Strait. Both routes were obstructed by large ice floes that precluded progress, and the mission failed. The expedition, however, gained for Franklin important experience in navigation of those icy seas.

The next mission, that of Sir John Ross in 1818, was specifically carried out to determine "the existence of a north-west passage from the Atlantic to the Pacific," as evidenced in the title of Ross's narrative that was published in 1819: *A Voyage of Discovery . . . in His Majesty's Ships Isabella and Alexander for the Purpose of Exploring Baffin's Bay and Inquiring into the Probability of a North-West Passage.* Ross rediscovered Baffin's Bay and reached as far west as Lancaster Sound, but he came to the erroneous conclusion that the sound was blocked to the west by a mountain range, which he named the Crocker Mountains, honoring the First Secretary of the Admiralty.

William Edward Parry, who conducted four expeditions to the Arctic between 1819 and 1827, made a major contribution to the definition of a Northwest Passage. During the first voyage that he led, in 1819–20, the two vessels, *Hecla* and *Griper,* under his command, entered Lancaster Sound, which Ross had explored the previous year. But Parry sailed westward beyond Ross's imaginary "Crocker Mountains," proving that they did not exist. Proceeding through what is now known as Parry Channel, the ships reached beyond 110° West Longitude. This feat satisfied the stipulation of the Parliamentary Act of 1818 authorizing the payment of £5,000 to those who passed 110° West Longitude to the north of America. The winter that the vessels spent in Winter Harbour off Melville Island, which they named, marked the first time a nineteenth-century seaborne expedition wintered in the Arctic. The introduction of canned food early in the century greatly facilitated the experience.

In August 1820, the vessels continued westward from Melville Island toward the Bering Strait, but were stopped by ice when they almost reached 113° 47' West Longitude. The accomplishment enforced a belief in the possibility of a Northwest Passage. From their most westerly location, the ships returned home to England via Davis Strait. Parry's successful expedition resulted in many names becoming attached to geographic features, including, predictably, Cape Franklin honoring Captain John Franklin.

Franklin's first appearance in the Arctic occurred on the 1818 exploration led by Captain David Buchan. Franklin's next contributions to the appreciation of the geography of the region occurred during the two overland expeditions that he led in 1819–22 and 1825–27. The first expedition was undertaken to map the North American coastline to the east and west of the mouths of the Coppermine and Mackenzie Rivers by determining the latitude and longitude of several points.

After crossing the Atlantic Ocean, Franklin's ship, the *Prince of Wales*, arrived at York Factory on the southwest corner of Hudson Bay. On September 9, 1819, Franklin and his exploratory party left York Factory, which would remain a Hudson's Bay Company post for 275 years until 1957, and traveled up the Hayes River, on to the northeast corner of Lake Winnipeg, and then to Cumberland House on the Saskatchewan River, a journey of about 700 miles. Franklin proceeded during the winter, traveling over 850 miles to Fort Chipewyan on Lake Athabaska.

In July 1820, the group left the fort and journeyed to Fort Providence on the north Shore of Great Slave Lake, and then continued overland to the Winter River where they established a winter's quarters, which they named "Fort Enterprise." In June 1821, the explorers left that fort, traveled to the Coppermine River, and followed the River in canoes about 334 miles to the point where it emptied into the Polar Sea. They noted that the mouth of the river was much farther south than Samuel Hearne had calculated, but still honored the explorer with the naming of Cape Hearne. Franklin's men explored approximately 500 miles of coast east of the Coppermine River, including Bathurst Inlet, but were thwarted in the effort to reach the west coast of Hudson Bay. The survivors of the journey, who endured sickness and starvation, reached their starting point at York Factory in July 1822. During the trek they were forced to eat mocassin leather, leading to Franklin's designation as "the man who ate his boots."

Interposed between Franklin's explorations, Captain William Edward Parry commanded his second expedition in search of the Northwest Passage in 1821–23. After crossing the ocean, the two vessels, the *Fury* and *Hecla*, sailed north of Southampton Island to the Frozen Strait. Coasting the shore line of Repulse Bay, the explorers laid to rest the concept of a westward passage from the Bay, thereby vindicating Captain Middleton, who had come to the same conclusion in 1742. During the travels, Parry named Fox's (Foxe) Channel and Baffin Island. The crews wintered at Winter Island, which they departed in July 1822, and explored and named the Melville Peninsula to the north. They entered a strait, named Fury and Hecla Strait for the two ships, but their progress west was aborted by ice. Once again, the ships wintered off Winter Island, and in August 1823 made a second attempt to sail through the strait, but again failed. The ships then returned to England. In 1824–25, Parry led his third and last attempt to traverse a Northwest Passage. His goal was to follow a different course, this time along Prince Regent Inlet, avoiding coastal regions. The mission was a dismal failure. Parry's voyages did contribute significantly, however, in that they opened the eastern entrance of the Northwest Passage and answered the question posed by Barrow: whether Greenland had a land connection with America that would close the Northwest Passage. No such connection exists.

In anticipation of Parry's successful completion of a Northwest Passage by sea and of Franklin's second overland expedition reaching the Pacific Ocean, the British Admiralty dispatched Captain F. W. Beechey, who had sailed with Franklin in 1818 and Parry in 1819–20, to meet the two expeditions. Beechey left England on the *Blossom* in April 1825 and rounded Cape Horn. Initially, he charted islands in the Pacific, including Pitcairn Island, where he met John Adams, the surviving mutineer from the *Bounty*. In June 1826, while in a Russian port, Beechey learned that Parry's mission had failed. But Beechey continued through Bering Strait, and then sailed along the north coast of

Alaska, leaving messages for Franklin's party. Thomas Elson, Master of the *Blossom*, was sent off with several men in a barge to explore the coast to the east. The group reached as far east as Point Barrow, which was named for Sir John Barrow of the Admiralty, thereby coming within 146 miles of Return Reef, which was the most westerly point reached by Franklin's overland journey. At that point in time, the charting of the Northwest Passage lacked only a small segment.

Franklin's second journey to the Polar Sea, in 1825–27, was a deliberate search for a Northwest Passage via an overland route. The expedition was facilitated by the coalition of the Hudson's Bay Company and the North West Company that occurred in 1821 after the beginning of Franklin's first journey. The plan was to proceed overland to the mouth of the Mackenzie River and continue westward by sea to the northwest corner of the continent which was, at the time, in Russia hands. The party set sail from Liverpool in 1825 and, after arriving in New York, continued on to a naval depot on Lake Huron where Franklin learned of the death of his wife.

The group arrived at the Hudson's Bay Company's first permanent western inland settlement, Cumberland House, located in current Saskatchewan. After making preparations to winter at Great Bear Lake, Franklin, accompanied by several of his group, descended the Mackenzie River to its mouth, where they landed, in August, on an island they named Garry Island. They returned to Great Bear Lake, where they met up with the remainder of the expedition. The winter residence site was named Fort Franklin in honor of the commander. In July 1826, the group left their wintering place and, after reaching the delta of the Mackenzie River, split into two expeditions. Dr. John Richardson led one group on a mission to survey the 500-mile coastline between the Mackenzie and Coppermine Rivers. Franklin commanded the other two small vessels, which traveled in a northwesterly direction. They ended the westward exploration at 149° 37' West Latitude and named the small area of land "Return Reef" before returning to the mouth of the Mackenzie River and to their wintering quarters on Great Bear Lake.

From July 4 to September 1, 1826, the two crews led by Dr. Richardson completed the survey of the coast between the Mackenzie and Coppermine Rivers. During the trip they named Cape Bathurst and Cape Parry on the shore of a bay to which they assigned Franklin's name. They named a headland Point De Witt Clinton for the Governor of New York, and a large island Wollaston Land, after the English philosopher, Dr. Hyde Wollaston. The latter name remains on a peninsula of what is now called Victoria Island. The strait between that island and the mainland was called Dolphin and

Union Strait, named for the expedition's small vessels. Richardson speculated that the Dolphin and Union Strait could constitute a portion of the Northwest Passage. Richardson proved to be prophetic, because Captain Richard Collison would sail through the strait eastward in 1852 and westward in 1853. Also, Raold Amundsen would cruise through that strait on his westward voyage in 1905, as would the St Roch in 1940–42 and again in a single season in 1944. After wintering at Fort Franklin, Franklin's group met up with Richardson's men at Fort Cumberland, and the expedition eventually reached England in September 1827.

Although Captain John Ross's reputation suffered greatly when the existence of the "Crocker Mountains" that he described on his first Arctic expedition in 1818 was disproved, he boldly proposed to the Admiralty in 1828 that a shallow draft steamship be used for an expedition in search for the Northwest Passage. The proposal was rejected, but Ross was able to gain financial support from Felix Booth, the Sheriff of London, who had made his fortune from the sale of Booth's gin. Booth would eventually be rewarded by the naming of Boothia Felix, the Gulf of Boothia, and Felix Harbour in the polar region. The expedition, on which John Ross was accompanied by his nephew, Commander James Clark Ross, was distinguished by the four consecutive winters, 1829–33, spent in the Arctic.

A paddle steamer, the *Victory,* left the Thames, and with a small sailing vessel in tow crossed Baffin's Bay, entered Lancaster Sound, and proceeded to Prince Regent Inlet. Near Fury Point, the crew encountered a tent and canned provisions from Captain Sir Edward Parry's *Fury* that had been wrecked four years earlier. The group wintered along the shore of the Gulf of Boothia southwest of the Fury and Hecla Strait. In 1830, James Clark Ross, sledging across the ice, discovered the island of King William's Land and named its northern point Cape Felix. During the trek, Ross also named Cape Abernathy for his second mate, and Cape Franklin for the Arctic explorer. In 1831, James Clark Ross, using several scientific instruments, defined the position of the North Magnetic Pole, which migrates over the years. The *Victory* had to be scrapped because it was no longer capable of sailing, and the crew returned to England on the *Isabella*, which was sailing in the vicinity of the 1832–33 wintering place.

Between 1836 and 1839, the maps of the northern coast of North America were more accurately extended and detailed. An expedition led by Peter Warren Dease, Chief Factor of the combined Hudson's Bay and North-West Company, and Thomas Simpson, a well-educated mathematician and surveyor, not only continued Franklin's earlier westward survey beyond Return Reef to Point

Barrow but also explored easterly to the east coast of Wollaston Island. The expedition had as its specific goal the completion of the survey of the northern coast of the continent by connecting surveys of Sir John Franklin from the east with those that had been made by Captain Beechey from the west. In July 1836, Peter Dease arrived at Fort Chipewyan on Lake Athabasca. In early December, Simpson left a settlement on the Red River and traveled over land about 1,300 miles to the lake, where he met up with his co-leader. After building two boats, the *Castor* and the *Pollux*, the expedition left the fort at the beginning of June 1837, and proceeded northward on the Great Slave Lake, and then downstream in the Mackenzie River. They eventually entered the Arctic Ocean where they were met by Inuit in kayaks. On July 23, the explorers reached Franklin's Return Reef. Simpson, accompanied by five men, continued in a westerly direction, initially on foot and subsequently in an umiak, a shallow draft skin boat provided by Inuit.

In early August 1837, Simpson reached Point Barrow, took possession of the land in Her Majesty's name, and, after taking measurements of the tides, returned to the mouth of the Mackenzie River. Simpson named the Colvile River after Andrew Colvile of the Hudson's Bay Company, and a small island offshore the mouth of the river, Esquimaux Island. The Fawn River near Return Reef was named for a fawn sighted there. The boats were towed upriver and continued up Bear Lake River, and across Great Bear Lake to the winter quarters on the shore of a river that was named for Peter Dease. In June 1838, Dease and Simpson left their winter quarters, which they had named Fort Confidence, and headed for the Coppermine River. Reaching the river in three weeks, the group proceeded downstream and, two weeks later, entered the open seas north of the continent. The party split, Dease remaining with the boats, and Simpson traveling overland to the Coppermine River. During his trek, Simpson passed Franklin's Point Turnagain and named Cape Franklin for the explorer, Cape Alexander for his brother, and the land to the north, Victoria Land (now Victoria Island) for the newly crowned queen. Simpson and his men met up with the Dease party at the mouth of the river, and the entire group returned to Fort Confidence to settle in for the winter. In 1838, a map published in the *Journal of the Royal Geographical Society* depicted the coastline between Return Reef and Point Barrow (see figure 35).

In 1839, the explorers once again descended the Coppermine River and exited into the Polar sea on the *Castor and Pollux*. They reached Cape Barrow, where they viewed Coronation Gulf; they then traveled east, naming Melbourne Island for the British Prime Minister, and reached Labyrinth Bay. They named the stream, which marked the farthest point that the boats reached, the Castor and Pollux River. In August, the party crossed Simpson Strait from Adelaide Peninsula, which they named for Queen Adelaide, the widow of King William IV. They sailed along the southern shore of King William Island and crossed from Melbourne Island to the southern shore of Victoria Island, from which point they crossed two bays that they named after the Duke of Cambridge and the Duke of Wellington. The channel separating these bays from the Kent Peninsula was named Dease Strait. The journey constituted the longest voyage made up to that time on the polar seas, about 1600 miles. Thomas Simpson's *Narrative of the Discoveries on the North Coast of America*, published posthumously in London in 1843, three years after Simpson died in a fight on the American prairie, gave testimony to the completion of the definition of the north coast of the North American continent. The possibility of a Northwest Passage was enhanced, but passage remained to be achieved.

In the chronicles regarding the search for a Northwest Passage, during the remainder of the nineteenth century, the name of Sir John Franklin was greatly amplified and reappeared often in the headlines of the press. It is ironic that the absence of the veteran Arctic explorer was ultimately responsible for the permanent presence of a complete Northwest Passage on maps. The uncharted segment of the passage was charted during the search for Franklin, which extended over three decades.

Shortly after he returned from his overland expedition of 1825–27, during which the north coast of America was charted from 149° 37' West Longitude at Return Reef, to the mouth of the Coppermine River, John Franklin proposed to the Admiralty an undertaking to complete the survey of the north coast of America, but the proposal was initially rejected. After serving in a variety of capacities in the Mediterranean Sea and Tasmania, Franklin was finally given command the expedition he wanted, the opportunity to complete the search for the Northwest Passage.

At the time that Franklin, who had been knighted in 1829, left on his third expedition in 1845, the only segment of a potential Northwest Passage that remained unexplored was a small area of seas, the northwest corner marked by Banks Land, the northeast corner by Cape Walker, the southwest by Wollaston Land, and the southeast by King William Land. If this region could be traversed, the long sought-for Northwest Passage would be established. As Sir John Barrow indicated in his proposal, the only portion of the Polar Sea that remained to be navigated was the 900 miles between the meridian of Melville Island and that of the Bering Strait. Franklin suggested that the passage could more likely be achieved by sailing between Cape Walker and Banks Land or, alternatively, by the Wellington Channel.

Two ships, the *Erebus* with the mission's commander aboard and the *Terror* with Captain Francis Rawdon Moira Crozier as second-in-command, were assigned for the mission. For the first time in the history of polar exploration, the two ships were equipped with auxiliary steam engines that propelled screws, which could be lifted out of the water when the ships were under sail. The vessels were equipped with ample provisions to extend an exploration over three years. The two ships departed England on May 19, 1845; they were last sighted of the coast of Greenland. The crew of 129 officers and men was never seen alive again by any European.

Lady Franklin, Sir John Franklin's second wife, first expressed fears concerning the status of the expedition in 1846 and, for more than a decade, would inspire search efforts. After several Arctic whalers returned from Davis Strait with no news of the whereabouts of the Franklin expedition, Sir John Richardson, a sixty-year-old physician at the Royal Naval Hospital who had accompanied Franklin on his first expedition, and Dr. John Rae as Richardson's second officer, left England in 1848 to relieve Franklin, who was thought to be detained in the Arctic. Richardson and Rae were to travel overland and down the Mackenzie River to the Arctic Ocean, and then east to the mouth of the Coppermine River. They planned to search along the coast of Wollaston Land. At about the same time, the *Plover* and the *Herald* undertook a mission to search along the coast of Alaska and as far east as the Mackenzie River. The two ships were joined by an English yacht, the *Nancy Dawson*, which rounded Cape Barrow, and thus became the first ship to sail in the Beaufort Sea. A third prong of the search, led by Sir James Clark Ross, was to approach from the Atlantic Ocean and cover the areas of Lancaster Sound, Barrow Strait, Wellington Channel, Cape Clarence, and Cape Walker. They were also to seek Franklin's party along the west coast of "Boothia Felix" and south of Prince Regent Inlet. None of these efforts met with success in uncovering evidence of Franklin's expedition, even though Ross sailed down Peel Sound through a strait that Franklin had traversed.

Many vessels joined in over forty search expeditions that were conducted in the first decade after Franklin's disappearance. Although they failed in their primary mission, they contributed significantly to improving the charts of the region. Evidence of Franklin's winter quarters of 1845–46 were discovered first by Captain Erasmus Ommanney. The information was brought back to England in the form of a message found on the north shore of Barrow Strait in 1850 by the crew of the *Prince Albert*, which was conducting a search sponsored by Lady Jane Franklin. The unpublished report stated that three graves were found on the northern part of Beechey Island. The gravestones spelled out the names and the death dates of three of the crew members, who died in the early part of 1846.

In 1850, an American expedition left New York to participate with the British in their search. Captains William Penny and Alexander Stewart, commanding the *Lady Franklin* and *Sophia*, investigated westward from Beechey Island. Their search ended without success, but Penny's conviction that Franklin's ships had sailed through Wellington Channel proved to be correct. Also in 1850, two naval ships, the *Enterprise*, commanded by Captain Richard Collinson, and the *Investigator*, led by Commander Robert McClure, left England for the Pacific Northwest. They missed their rendezvous in the Hawaiian islands and proceeded to the Bering Strait and the Beaufort Sea independently.

After entering the Beaufort Sea, the *Investigator* followed the coast that Franklin and Dease and Simpson had previously traversed. In the process, McClure discovered Prince of Wales Strait between Banks Island and Victoria Island. The ship's progress was stopped by ice, and it was obliged to end its easterly progress near the northwestern end of Prince of Wales Strait. But, as the crew looked across to Melville Island sixty miles away, they could see the last link of the Northwest Passage to be navigated.

During McClure's search for Sir John Franklin from 1850 through 1854, he found two routes for a potential Northwest Passage. One passed through Prince of Wales Strait while the other coursed north of Banks Island through what later became McClure Strait. After spending two winters in the Arctic, the crew abandoned the *Investigator* and returned on the *Resolute,* commanded by Captain Henry Kellett, which had been dispatched by the Admiralty in 1852 to deposit provisions on Melville Island as part of the search for Franklin. McClure is credited as being the first to connect the previously explored eastern and western ends of the Northwest Passage. McClure completed the Northwest Passage from west to east, although this required three different ships over water and a sled over one land segment. As a result, Parliament awarded £10,000 to the crew of the *Investigator* for the discovery of the Northwest Passage. Thus, by 1852, all parts of the passage had been defined, but over a half-century would pass before any vessel would complete the passage.

Captain Collinson, McClure's senior officer, spent the winter of 1850 in Hong Kong, and the following year, after rounding Point Barrow from the west, sailed up Prince of Wales Strait, where the

Enterprise wintered. In 1854, the expedition proved that Wollaston, Prince Albert, and Victoria were one landmass. The ship traversed Dolphin and Union Strait, Coronation Gulf, and Dease Strait, and wintered in Cambridge Bay. In August 1854, the *Enterprise* finally sailed around Point Barrow from the east after spending a total of 1164 days in the Arctic.

In 1852, while searching for Franklin, Captain William Kennedy, aboard the *Prince Albert,* discovered the narrow Bellot Strait (later named for Joseph René Bellot, who was second-in-command on that voyage and later drowned during the Belcher expedition) between North Somerset and the Boothia Peninsula. This strait would become an integral segment of the Northwest Passage. From 1852 through 1853, Sir Edward Belcher commanded a search squadron consisting of five naval ships. After the crossing from Greenland, the vessels joined up at Beechey Island but failed in their search for signs of Franklin's ships. Two of the vessels sailed into Queen's Channel and then discovered Northumberland Sound, named after the Duke who presided over the Board of Admiralty. During the voyage, Belcher discovered Princess Royal Islands. Commander Francis Leopold McClintock, aboard the steam-propelled *Intrepid*, as part of Belcher's squadron, discovered Prince Patrick Island to the north. Additional probing of the Arctic, as part of the search for Franklin's relics, was conducted aboard ships and sleds in 1853–55. Included among these was an expedition led by Dr. Elisha Kent Kane that including ten members of the United States Navy.

John Rae, in October 1854, issued his summary to the Admiralty, detailing his discovery of relics from the Franklin expedition. At the time, he presented spoons and forks bearing the initials of officers of the two vessels involved in the voyage, and a gold watch engraved with the name of the ice master of the *Erebus*, James Reid. As a consequence, the British government honored its previous offer and rewarded Rae and his men £10,000 for determining the fate of the Franklin expedition.

Rae, a physician, became a Chief Factor in the Hudson's Bay Company in a unique way and contributed significantly to the discovery of the Northwest Passage and to an appreciation of the geography of the Arctic. A graduate of the Edinburgh Medical College, Rae began his relationship with the Hudson's Bay Company as post physician at Moose Factory on James Bay. In 1846, he set forth to survey and map the northern shores of the continent from the Straits of Fury and Hecla, thereby completing surveys made for the Company by Thomas Simpson and Peter Warren Dease. The object was to determine whether or not Boothia was a peninsula and also whether

King William was an island, both essential elements in the definition of a potential Northwest Passage. Rae and twelve men sailed northward from Churchill on the west shore of Hudson Bay to Repulse Bay where they wintered. From that point, in the course of a half-dozen land journeys north and west, Rae charted 655 miles of previously unmapped coastline and proved that Boothia was a peninsula, which precluded a Northwest Passage south of the 68th parallel.

In 1848, as part of the three-pronged search for Franklin along the north coast of America from Mackenzie River to Coppermine River, Rae, accompanied by Dr. John Richardson, failed to find any traces of the lost ships or its crew. After Richardson returned to England, Rae remained, and, in 1851, crossed over the ice of the Dolphin and Union Strait to the Wollaston Peninsula of Victoria Island. He proceeded along the south coast as far west as Prince Albert Sound, and then followed the south shore of Coronation Gulf and Dease Strait to the northern tip of Kent Island. Rae and his men crossed Dease Strait to survey the southern coast of Victoria Island. In all, his travels on ice of 1,080 miles, coupled with 1,390 miles on water, defined 630 miles of previously uncharted coasts. During these explorations, which earned him the Royal Geographical Society's Gold Medal, Rae unknowingly passed within forty miles of Franklin's abandoned ships.

Rae's discovery of the relics of the Franklin expedition occurred during his fourth Arctic expedition in 1853–54, which was conducted not as a search mission, but rather to define the geography of Boothia. Rae left York Factory and proceeded north by boat to Repulse Bay, where six remaining members of the party wintered. In the spring of 1854, the group set out north for the Castor and Pollux River. After discovering a river that he named Murchison after the President of the Royal Geographical Society, the group crossed a bay, which they named Shepherd Bay after the Deputy Director of the Hudson's Bay Company. They also named a small island Bence Jones after the distinguished physician who prepared an extract of tea for the voyage. Rae surveyed the land from King William Land to Repulse Bay and determined that King William Land was separated from Boothia by an open waterway through which Roald Amundsen cruised on the first navigation of the entire Northwest Passage.

On the outward journey, Rae met a Native, In-nook-poo-zhee-zook, who was wearing a gold braid naval cap band and told of about forty white men who had starved to death about a ten days' journey to the west. When Rae returned to Repulse Bay, a group of Inuit completed the tale, indicating that they had initially encountered the white men, in 1850, dragging a boat over the ice near King William Land.

Later that year, the Inuit discovered about forty corpses, some of which were in tents or under a boat. Rae incorporated in his report that, according to the narratives, some of the Franklin party had resorted to "cannibalism as a means of sustaining life." The cannibalism was authenticated in the 1980s by studies of bones found preserved by the ice, demonstrating that they had been sawed off parts of the bodies.

Dr. John Rae, whose journeys over the year encompassed 2,300 miles, has his name perpetuated in the Arctic on Rae Strait, Rae Isthmus on the Melville Peninsula, and Rae River, which exits into Coronation Gulf.

In May 1845, Captain Sir John Franklin had sailed down the Thames on the last expedition sent by the Admiralty in search of the Northwest Passage. A decade later, in 1855, an overland expedition led by James Anderson, a Chief Factor in the Hudson's Bay Company, and James Green Stewart ended the British government's involvement in the dramatic search for the remains of Franklin and his crew. The Admiralty's last undertaking was an attempt to expand on the discovery by Dr. John Rae of the first relics of the Franklin expedition and to assess the validity of Rae's official report. Rae's report was authenticated by conversations with the Inuit, and Anderson and Stewart's expedition found a boat plank bearing the name of one of Franklin's vessels and a snowshoe inscribed with the name of the ship's surgeon.

Lady Franklin's continued dedication to the search for her husband and, short of finding him, to determining his accomplishments on the journey led to an expedition commanded by Captain Francis Leopold McClintock. The search group left England in June 1857, on the *Fox,* a schooner with an auxiliary steam engine, and, after wintering off the west coast of Greenland, sailed across the ocean to Beechey Island, from which they followed Prince Regent Inlet into Bellot Strait.

McClintock named a wide strait after Sir John Franklin, and wintered at the eastern end of Bellot Strait. During the winter, McClintock led a group of men to the magnetic pole, and, in the course of their travels, traded with Inuit for relics of the Franklin expedition. The Inuit told of a ship that had been crushed by ice, but indicated that all the crew landed safely in the region of King William Island. McClintock's journey was particularly significant because it covered about 120 miles of coastline, and completed the survey of the continent's northern limits. The mapmakers could finally reinsert the previously deleted Northwest Passage, albeit not in its originally proposed location, and with a course that was more complex than envisioned.

In 1859, McClintock found relics of Franklin's party on Cape Felix on the west coast of King William Island. Two records of Franklin's expedition were found in cairns, one, in the hand of "James Fitzjames, Captain HMS *Erebus*," stated that the ships were deserted on April 22, 1848, and that Sir John Franklin had died of the June 11, 1847. Reconstruction of the course of the Franklin expedition indicated that the ships probably sailed up Wellington Channel and then through what became named Crozier Strait to the west of Cornwallis Island. In the summer of 1846, they sailed south from Peel Sound into Franklin Strait into open seas near King William Island. The ships never escaped the ice in Victoria Strait.

McClintock's expedition continued overland to the western part of King William Island, where the explorers found a large boat with clothing, guns, and the remains of two men. The expedition established the fate of Sir John Franklin, explored 380 miles, surveyed 800 miles of new coastline, and, most importantly, asserted that the Franklin Expedition was the truly first to discover a Northwest Passage.

Americans played roles in the epilogue to the discoveries of the English searches. Captain C. F. Hall spent ten years of his life, during the 1860s, searching for survivors and relics of the Franklin expedition. During his winters in the Arctic, he discovered relics of Frobisher's sixteenth-century expeditions; he also collected relics of the Franklin expedition and recorded Inuit narratives of their recollections concerning the fate of the members of that expedition. A skeleton, believed to be that of Lieutenant Le Vesconte of the *Erebus*, was dug up and later interred at Greenwich, England. In 1878–80, Lieutenant Frederick Schatka of the United States Army led a search for Franklin's relics. On King William Island, they uncovered more relics and bodies of the Franklin expedition.

The consequence of Franklin's expedition and the searches that ensued was that all the Arctic waterways could finally be included on maps. The wish for a Northwest Passage had come true. Piecemeal segments had been joined, just as in a jigsaw puzzle, and all that remained was for the navigation of that passage to be completed (see figure 36).

That historic accomplishment was achieved by Raold Amundsen aboard the *Gjoa*, a 47-ton herring boat with sails and a 13-horsepower engine, in a voyage that extended from 1903 to 1907. The 29-year-old Amundsen was accompanied by six men and several dogs. The group made landfall in the Canadian Arctic on Beechey Island. They then sailed down Peel Sound, passed Bellot Strait, and anchored in a harbor along the south coast of King William Island. At this point, the journey was interrupted to send sledding parties to the Magnetic North Pole, which Amundsen actually failed to reach. The ship resumed its voyage through the Northwest Passage in August 1905. When it reached Cape Colborne at the entrance to Cambridge Bay, the *Gjoa* had sailed through a segment of the waterway that had not been traversed previously. The ship proceeded through Dease Strait, Coronation Gulf, and Dolphin and Union Strait into open seas. This brought the travelers off Nelson Head, the southern Cape of Banks Island.

There they sighted the *Charles Hanson*, a schooner from San Francisco, thereby affirming that Amundsen's group had completed a Northwest Passage. The *Gjoa* continued west to the mouth of the Mackenzie River, where its progress was halted by ice, forcing the expedition to winter at that point. The voyage was resumed in July 1906, and, mainly with the power generated by the engine, the ship rounded Point Barrow and passed through the Bering Strait at the end of August, completing the first total navigation of the Northwest Passage. The crew was initially welcomed in Nome, Alaska, and arrived at San Francisco on October 19, 1906. The *Gjoa* remained in San Francisco until she returned to Oslo in 1972. Amundsen added to his fame in 1911 when he became the first person to reach the South Pole.

In 1908, Joseph Elzéar Bernier voyaged to Melville Island, where he spent two winters. He suggested in his log that a more northerly passage than the one taken by Admundsen would prove to be easier. The next complete passage was accomplished in 1940–42 by the Royal Canadian Mounted Police schooner, *St Roch*. That journey began in June at Vancouver, British Columbia. The group wintered in Walker Bay and proceeded eastward in July 1941. The route took them through Simpson Strait, around the southeast point of King William Island, and north through Rae Strait to the west coast of Boothia, where progress was stopped for the winter. In early August 1942, the voyage was resumed and the ship passed through Bellot Strait. On the eastern end of the strait, the *St Roch* stopped at Fort Ross, then proceeded to Pond Inlet on the east coast of Baffin Island, and reached Halifax on October 11, 1942. This constituted the first navigation of the Northwest Passage from the Pacific to the Atlantic.

In 1944, the *St Roch* became the first vessel to navigate the Northwest Passage from east to west in a single season. This time the route followed the northern passage through Parry Channel to Melville Island and across McClure Strait to Banks Island, as originally suggested by Joseph Elzéar Bernier in 1908. The voyage began at Halifax on July 22, 1944, and ended on October 16, 1944, at Vancouver, after a distance of 7,295 nautical miles.

In 1960, the USS *Seadragon* made the first transit by a submarine. The first commercial voyage was made in 1969 by the ice-breaking tanker SS *Manhattan.* In 1977, Willy de Roos of Holland single-handedly completed a passage on a thirteen-meter ketch. The MV *Lindblad Explorer* in 1984 became the first passenger ship to transit the Northwest Passage. Four years later, the same ship, renamed *Society Explorer*, made the transit in both directions. In the summer of the new millennium, the *St Roch II* traversed the Northwest Passage in one month without running into any ice. On that voyage, the crew came across the bodies of six of members of the expedition led by Sir John Franklin in 1845–48.

Cartographically, the Northwest Passage was on again, off again, and on again. If the recent voyage of the *St Roch II* is evidence, could it be that some day, in the distant or not-too-distant future, global warming will result in the fulfillment of the original wish for a more readily usable and commercially advantageous Northwest Passage?

References

Bernier, J. E. *Master Mariner and Arctic Explorer: A Narrative of Sixty Years at Sea from the Logs and Yarns of Captain J. E. Bernier.* Ottawa: Le Droit, 1939.

Best, George. *A True Discourse of the Late Voyages of Discoverie for the Finding of a Passage to Cathaya, by the Northwest under the Conduct of Martin Frobisher, Generall.* London, 1578.

Briggs, Henry. "Treatise of the North-West Passage to the South Sea, through the Continent of Virginia, and by Fretum Hudson." In Samuel Purchas, *Purchas His Pilgrimes.* London, 1625.

Charters, Statutes, Orders in Council, &c, Relating to the Hudson's Bay Company. London: The Hudson's Bay Company, 1949.

Davis, John. *The Worldes Hydrographicall Discription.* Sandrudg, 1595.

Dobbs, Arthur. *An Account of the Countries Adjoining to Hudson's Bay in the North-West Part of America.* London, 1744.

———. Communication in the *Monthly Miscellany or Memoirs for the Curious,* London, 1708.

Foxe, Luke. *North-West Foxe.* London, 1635.

Gerritz, Hessel, Pedro Fernández de Quirós, and Isaac Massa. *Descriptio ac delineatio geographica detectionis freti.* Amsterdam, 1612.

Golder, Frank A. *Bering's Voyages: An Account of the Efforts of the Russians to Determine the Relation of Asia and America.* 2 vols. New York: American Geographical Society, 1922.

Hakluyt, Richard. *Divers Voyages Touching the Discovery of America.* London, 1582.

Hearne, Samuel. *A Journey from Prince of Wales's Fort in Hudson's Bay to the Northern Ocean*. London, 1795.

Herrera y Tordesillas, Antonio de. *Novus orbis, sive Descriptio Indiae Occidentalis*. Amsterdam, 1622. Translated into German as *Zwölfter Theil der newen Welt, das ist: Gründliche volkommene Entdeckung aller der west indianischen Landschafften, Insuln*, . . . Frankfurt, 1623.

James, Thomas. *The Strange and Dangerous Voyage of Captaine Thomas James*. London, 1633.

Janes, John. Article, pp. 771–92. In Richard Hakluyt, *The Principall Navigations, Voyages, Traffiques, and Discoveries of the English Nation*. London, 1589.

Jode, Cornelis de. *Speculum Orbis Terrae*. Antwerp, 1593.

Markham, Clements R. *The Voyages of William Baffin, 1612–1622*. London, Hakluyt Society, 1881.

Mirsky, Jeannette. *To the Arctic! The Story of Northern Exploration from Earliest Times to the Present*. London: H. Hamilton, 1934.

Munk, Jens. *Navigatio septentrionalis*. Copenhagen, 1624.

Ramusio, Giovanni Battista. *Terzo Volume delle Navigationi et Viaggi*. Venice, 1556.

Ross, John. *A Voyage of Discovery, Made under the Orders of the Admiralty, in His Majesty's Ships Isabella and Alexander for the Purpose of Exploring Baffin's Bay and Inquiring into the Probability of a North-West Passage*. London: J. Murray, 1819.

Simpson, Thomas. *Narrative of the Discoveries on the North Coast of America: Effected by the Officers of the Hudson's Bay Company during the Years 1836–39*. London: R. Bentley, 1843.

Skelton, R. A. *Explorers' Maps*. London: Routledge and Kegan Paul, 1958.

Vancouver, George. *A Voyage of Discovery to the North Pacific Ocean and Round the World, 1791–1795*. Edited by W. Kaye Lamb. London: Hakluyt Society, 1984.

Waldseemüller, Martin. *Cosmographiae Introductio, cum quibusdam geometriae ac astronomiae principiis ad eam rem necessariis Insuper quatuor Americi Vespucij navigationes. Universalis cosmographiae descriptio tam in solido quam plano. eis etiam insertis, quae Ptholomaeo ignota a nuperis reperta sunt*. Saint-Dié, 1507.

Wytfliet, Cornelis van. *Descriptionis Ptolemaicae Augmentum*. Louvain, 1597.

SELECTED READINGS

Amundsen, Roald. *Roald Amundsen: My Life As an Explorer*. Garden City, N.Y.: Doubleday, Doran, 1928.

Beattie, Owen. *Frozen in Time: Unlocking the Secrets of the Franklin Expedition*. New York: Dutton, 1988.

Berton, Pierre. *The Arctic Grail: The Quest for the North West Passage and the North Pole, 1818–1909*. New York: Viking, 1988.

Christy, Miller. *The Voyages of Captain Luke Foxe of Hull, and Captain Thomas James of Bristol, in Search of a Northwest Passage, in 1631–32; with Narratives of the Earlier Northwest Voyages of Frobisher, Davis, Weymouth, Hall, Knight, Hudson, Button, Gibbons, Bylot, Baffin, Hawkridge, and Others.* London: Printed for the Hakluyt Society, 1894.

Cooke, Alan, and Clive Holland. *The Exploration of Northern Canada, 500–1920.* Toronto: Arctic History Press, 1978.

Crouse, Nellis Maynard. *The Search for the Northwest Passage.* New York: Columbia University Press, 1934.

Cyriax, Richard J. *Sir John Franklin's Last Expedition.* London: Methuen & Co., 1939.

Hakluyt, Richard. *The Principal Navigations Voyages Traffiques & Discoveries of the English Nation . . . within the Compass of these 1600 Years.* London: Dent, 1927.

Larsen, Henry A. *The North-West Passage, 1940–1942 and 1944: The Famous Voyages of the Royal Canadian Mounted Police Schooner St Roch.* Ottawa: Queen's Printer, 1958.

Newman, Peter C. *Company of Adventurers.* Vol. 1. Ottawa: Penguin Books, 1983.

Quinn, David B. "The Northest Passage in Theory and Practice." In *North American Exploration.* Edited by John Logan Allen. Vol. 1, *A New World Disclosed.* Lincoln: University of Nebraska Press, 1997.

Savours, Ann. *The Search for the North West Passage.* New York: St. Martin's Press, 1999.

Skelton, R. A. *Explorers' Maps.* London: Routledge and Kegan Paul, 1958.

Stefansson, Vilhjalmur. *Northwest to Fortune: The Search of Western Man for a Commercially Practical Route to the Far East.* New York: Duell, Sloan, and Pearce, 1958.

Thompson, George Malcolm. *The Search for the North-West Passage.* New York: Macmillan, 1975.

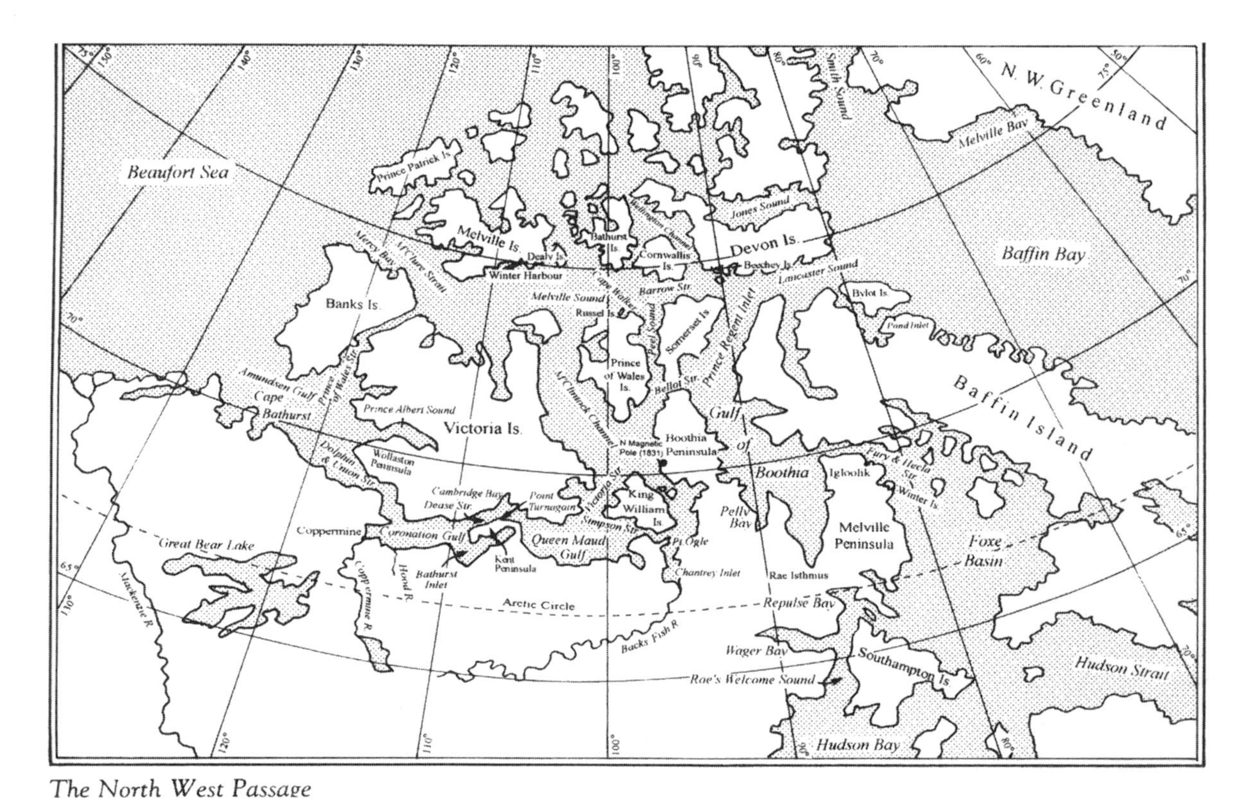

Fig. 21. "The Northwest Passage." From Ann Savours, *The Search for the North West Passage* (New York: St. Martin's Press, 1999); reproduced with permission.

Fig. 22. Paolo Forlani. "IL Disegno del discoperto della noua Franza' il quale s'è hauuto ultimamente dalla nouissima nauigatione dè Franzesi . . ." Venice, 1566. Copperplate, 280 x 400 mm. Private Collection.

The earliest printed map devoted to North America. The first map to show and name the Strait of Anian, depicted as flowing from "Mare Setentrionale In Cognito" (north of the North American continent) to "Golfo Chinan," which is a part of the Sea of China.

Fig. 23. Abraham Ortelius. "AMERICAE SIVE NOVI ORBIS, NOVA DESCRIPTIO." Copperplate, 370 x 500 mm. From Ortelius, *Theatrum Orbis Terrarum* (Antwerp, 1570). Private Collection.

The first map of the Americas to appear in a modern atlas. The northern part of North America is cut off by the margin, thus avoiding the issue of the presence of a Northwest Passage.

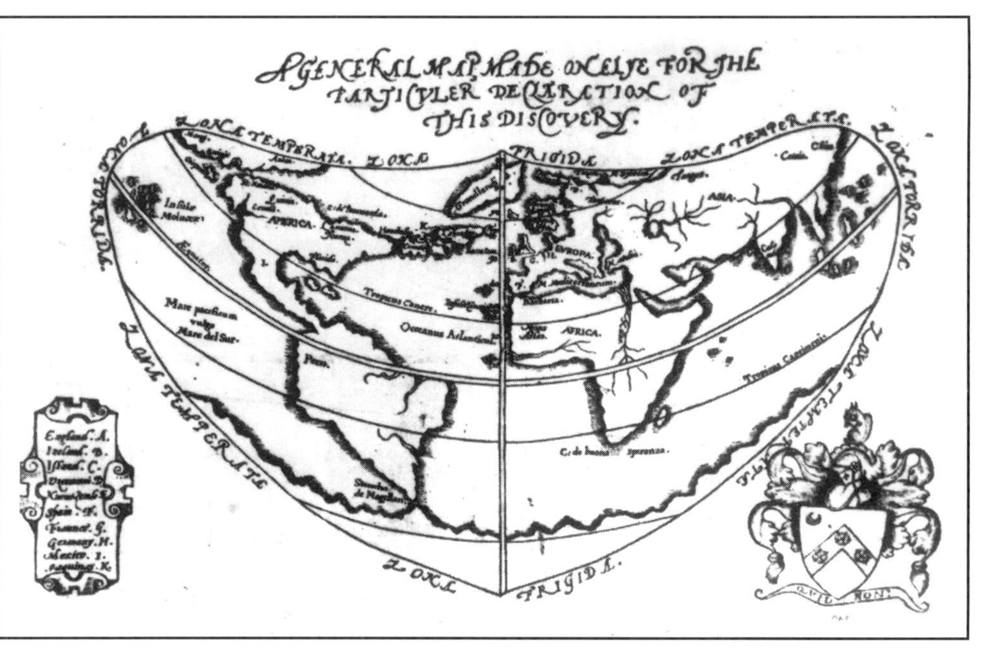

Fig. 24. Sir Humphrey Gilbert. "A General Map, Made Onelye for the Particvler Declaration of This Discovery." Engraving, 230 x 330 mm. From Gilbert, *A Discourse of a Discoverie for a New Passage to Cataia* (London, 1576). Photograph courtesy of John Carter Brown Library, Brown University, Providence, R.I.

The map indicates that by sailing between Greenland and Labrador, and via a Northwest Passage, shown as open water, a ship would reach the isles of the Moluccas.

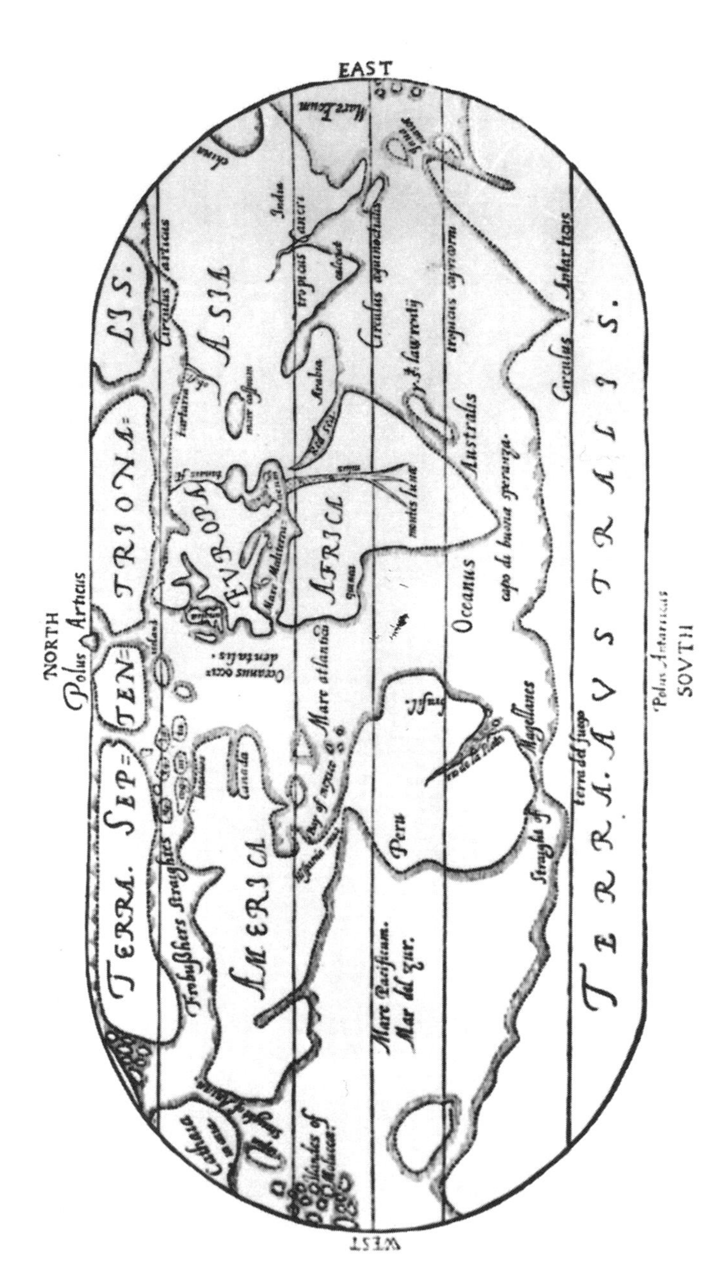

Fig. 25. George Best. [No Title]. World Map. Woodcut, 215 x 395 mm. From Best, *A True Discourse of the Late Voyages of Discoverie for the Finding of a Passage to Cathaya, by the Northwest, under the Conduct of Martin Frobisher, Generall* (London, 1578). Photograph courtesy of John Carter Brown Library, Brown Uiversity, Providence, R.I.

On the map there is a broad channel name "Frobusshers Straightes" connecting the Atlantic Ocean to the Straight of Anain and Cathaia and the "Ilandes of Molluca."

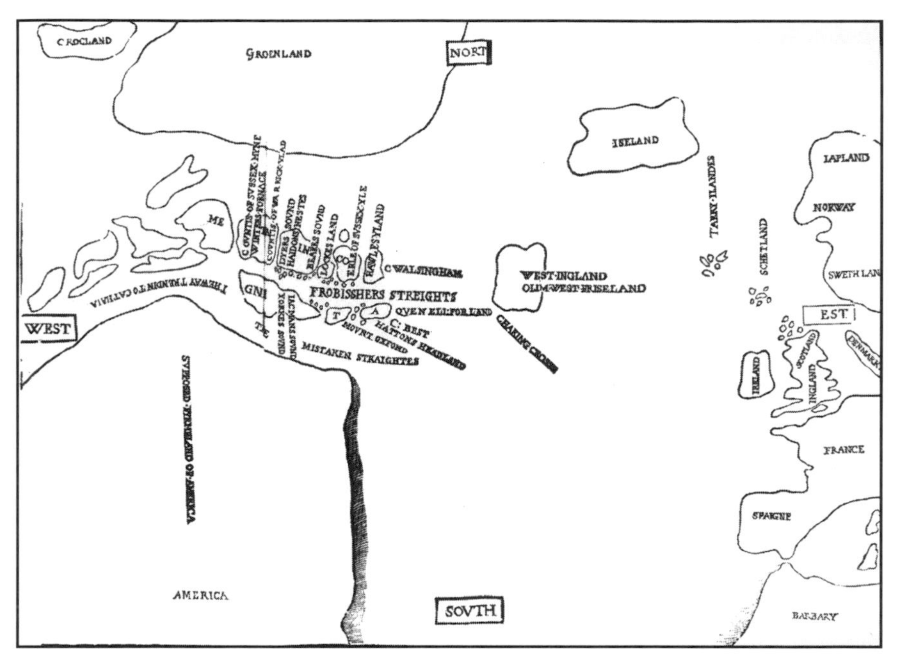

Fig. 26. George Best and James Beare. [No Title]. Woodcut, 285 x 395 mm. From Best, *A True Discourse of the Late Voyages of Discoverie for the Finding of a Passage to Cathaya, by the Northwest, under the Conduct of Martin Frobisher, Generall* (London, 1578). Photograph courtesy of John Carter Brown Library, Brown University, Providence, R.I.

The map accompanied the first account of the three voyages of Martin Frobisher. It depicts "Frobissher's Streights, "Mistaken Streightes" and several islands north of America.

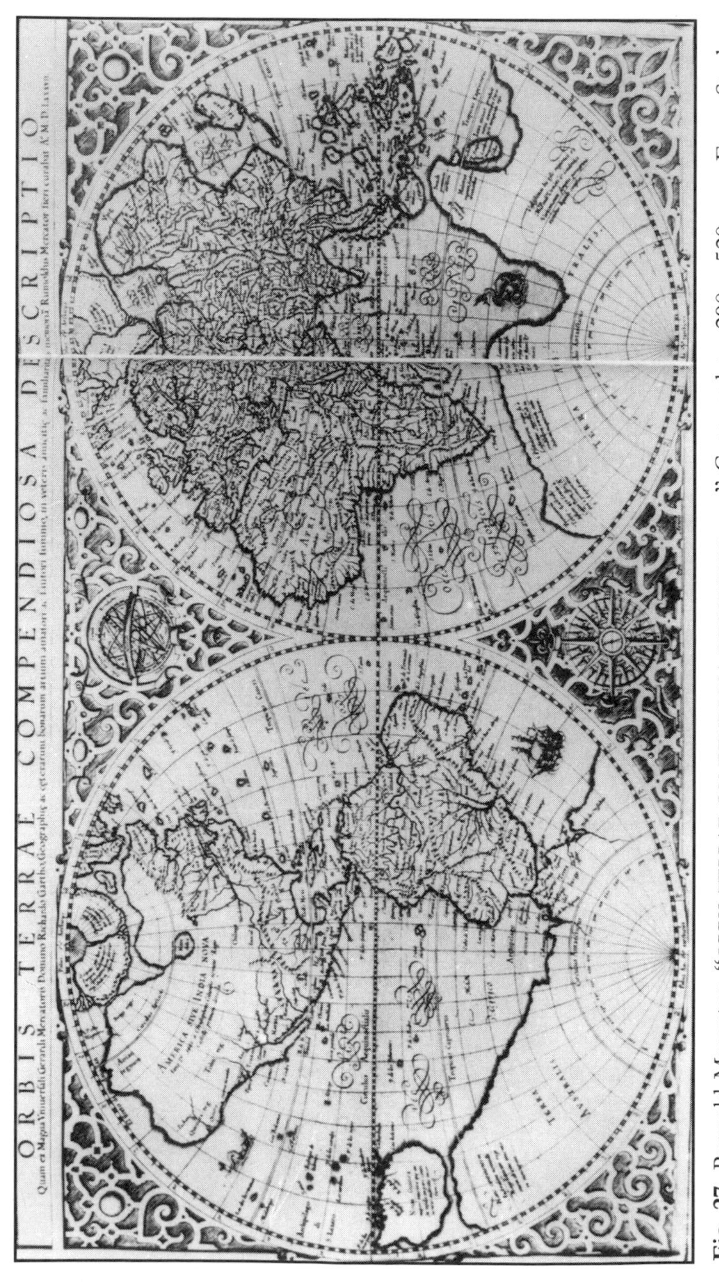

Fig. 27. Rumold Mercator. "ORBIS TERRAE COMPENDIOSA DESCRIPTIO . . ." Copperplate, 290 x 520 mm. From Strabo, *Strabonis Rerum Geographicarum* (Geneva, 1587). Private Collection. Depiction of an uncomplicated Northwest Passage.

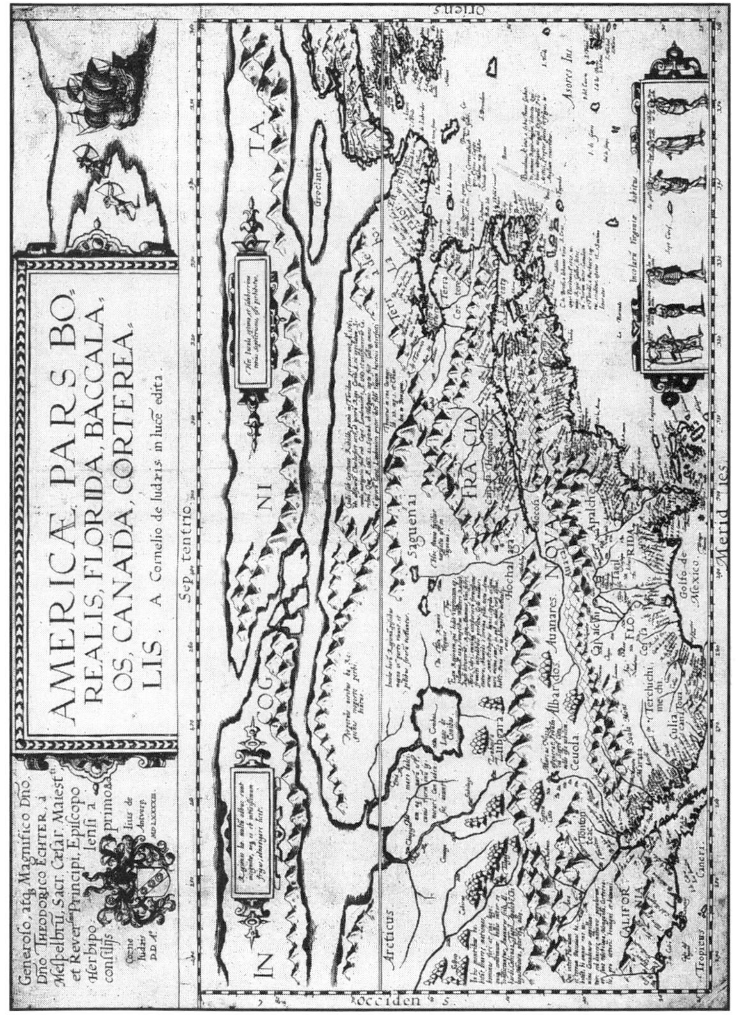

Fig. 28. Cornelis de Jode. "AMERICAE PARS BOREALIS, FLORIDA, BACCALAOS, CANADA, CORTEREALIS." Copperplate, 365 x 500 mm. From Cornelis de Jode and Gerard de Jode, *Speculum Orbis Terrarum* (Antwerp, 1593). Private Collection. A direct Northwest Passage is shown.

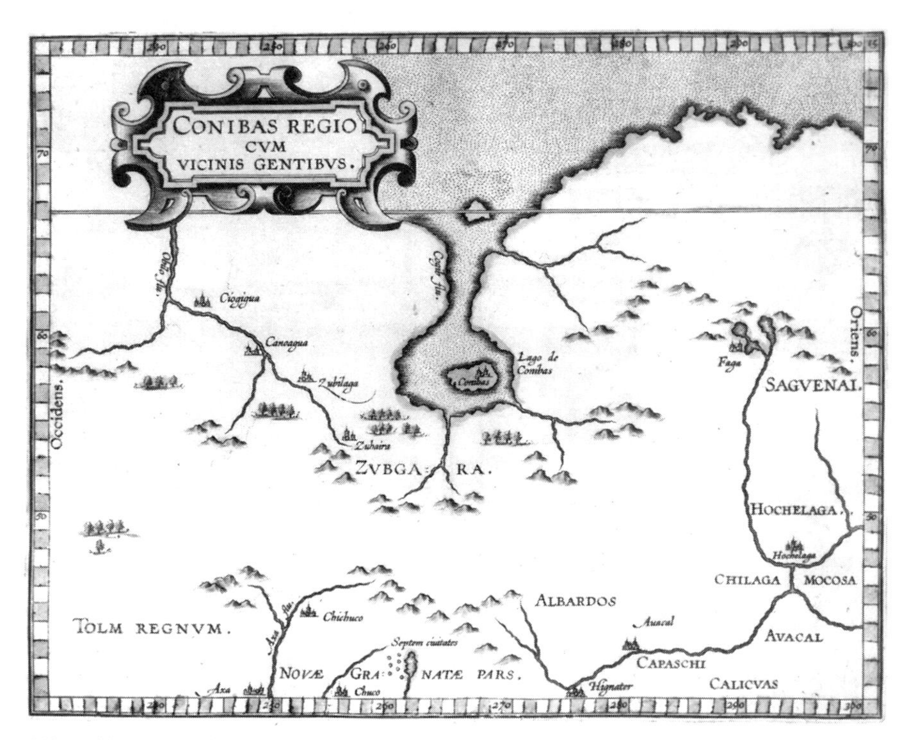

Fig. 29. Cornelis van Wytfliet: "CONIBAS REGIO CVM VINCINIS GENTIBVS." Copperplate, 220 x 270 mm. From Wytfliet, *Descriptiones Ptolemaicae Augmentum* (Louvain, 1597). Private Collection.

The map is contained in the first atlas devoted solely to the Americas. At the top of the map is the Northwest Passage, into which a "Lago de Conibas" empties.

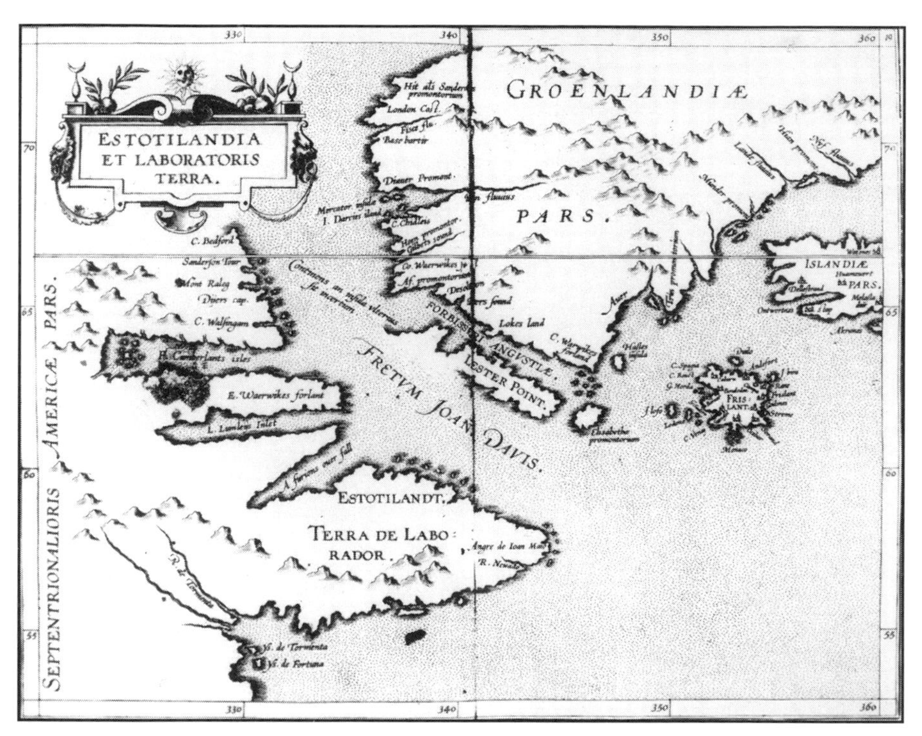

Fig. 30. Cornelis van Wytfliet: "ESTOTILANDIA ET LABORATORIS TERRA." Copperplate, 230 x 290 mm. From his *Descriptiones Ptolemaicae Augmentum* (Louvain, 1597). Private Collection.

The map concentrates on the area of the voyages of Frobisher and Davis in search of the Northwest Passage.

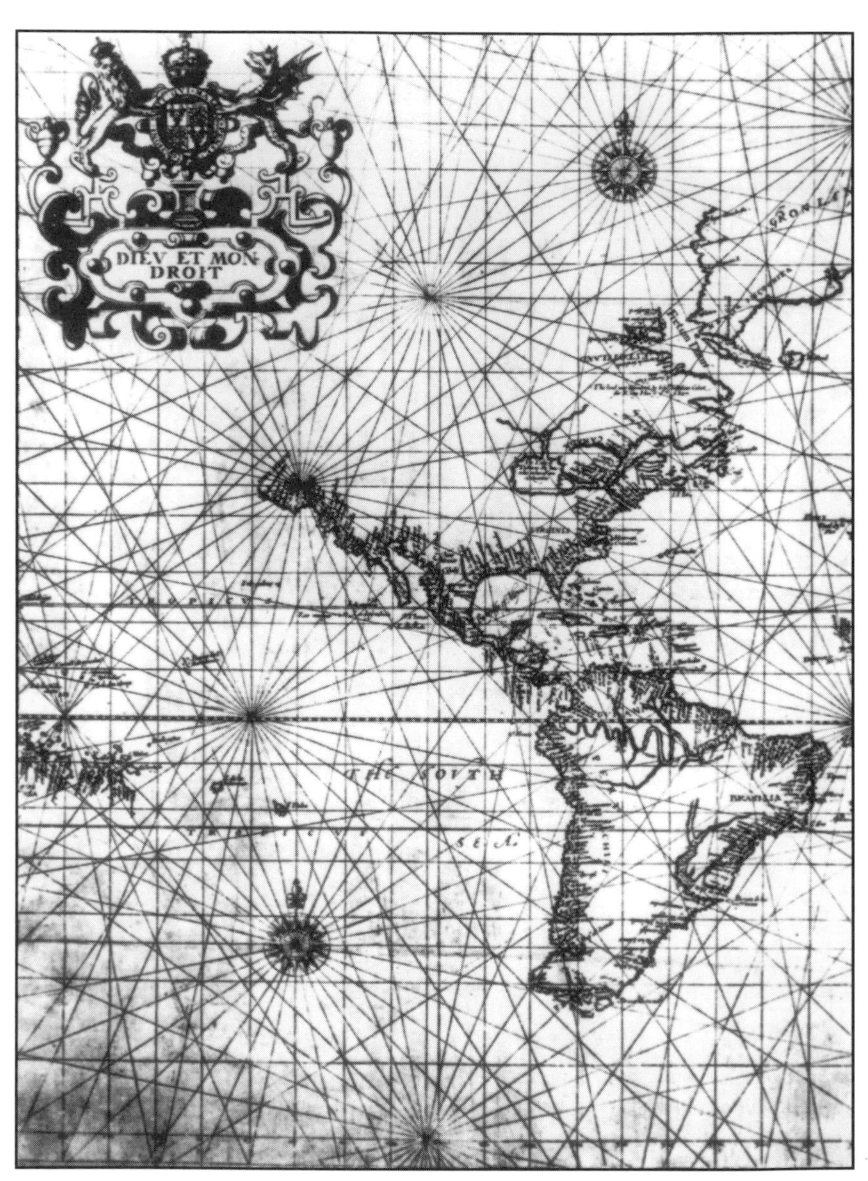

Fig. 31. Edward Wright and Emeric Molyneux. [No Title]. Copperplate, 430 x 640 mm. Detail from Richard Hakluyt, *Principal Navigations, Voyages, Traffiques and Discoveries of the English Nation* . . . (London, 1599). Private Collection.

The northern border of North America from the middle of the continent to the west coast is left blank.

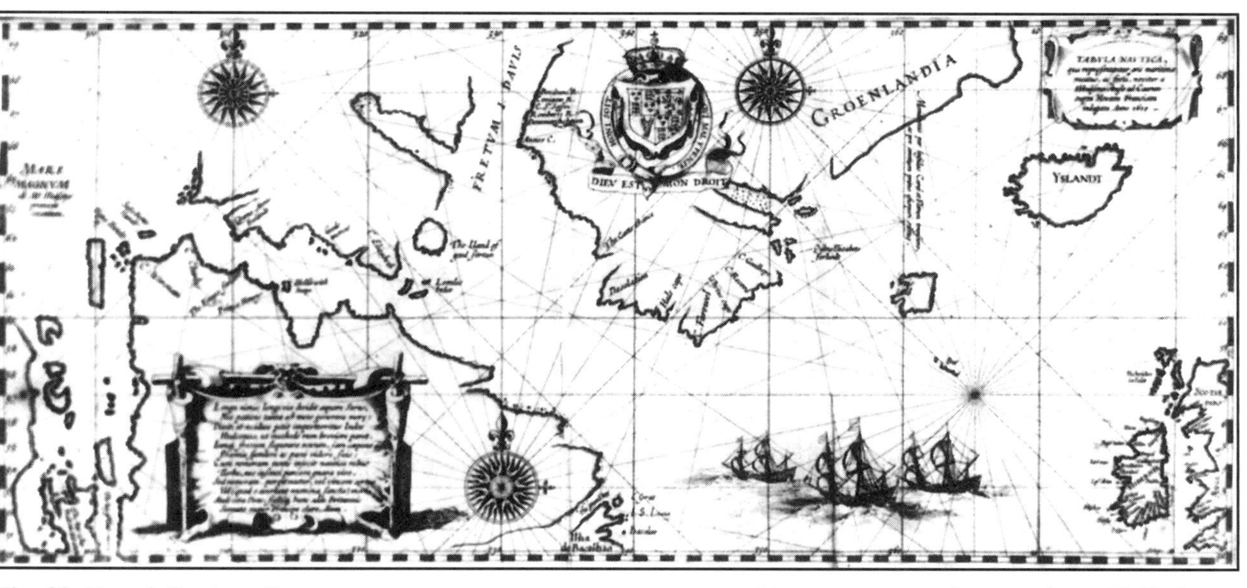

Fig. 32. Hessel Gerritsz. "TABVLA NAVTICA, qua repræsentantur oræ maritimæ. meatus, ac freta, noviter a H Hudsono Anglo ad Caurum supra Novam Franciam indigata Anno 1612." Copperplate, 240 x 520 mm. From Gerritsz, *Descriptio ac Delineato Geographica Detectionis* (Amsterdam, 1612). Private Collection.

The map depicts Hudson Strait emptying into Hudson Bay and indicates Hudson's conviction that he had completed a Northwest Passage by showing a large sea at the western boundary of the map.

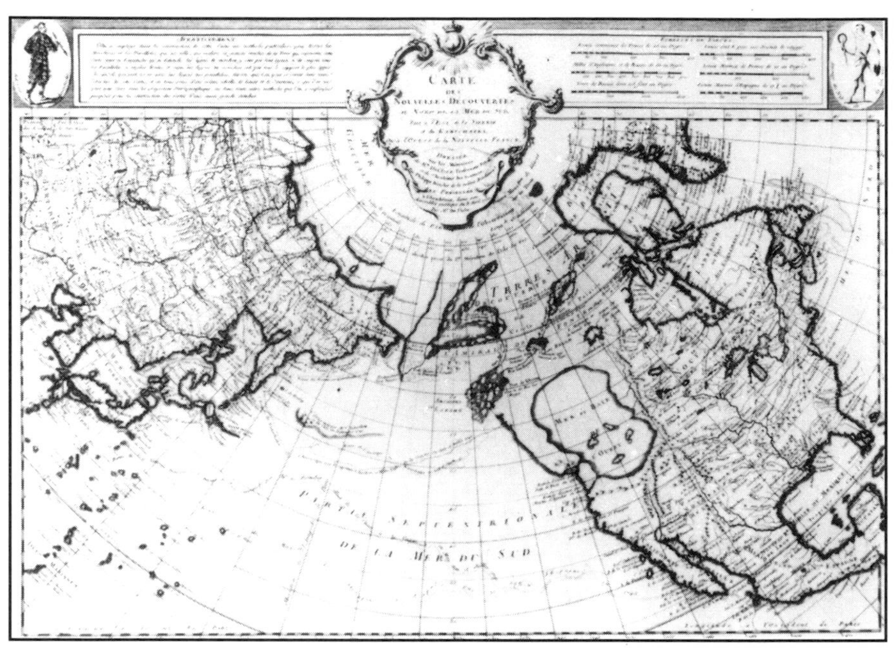

Fig. 33a. Philippe Bauche. "CARTE DES NOUVELLES DECOUVERTES au Nord de la Mer du Sud." Paris, 1752. Copperplate, 460 x 630 mm. Private Collection.

The map depicts Bering's two voyages for Russia. A fictitious 1200-mile-long "Mer ou Baye de l'Ouest" in northwestern America is shown for the first time.

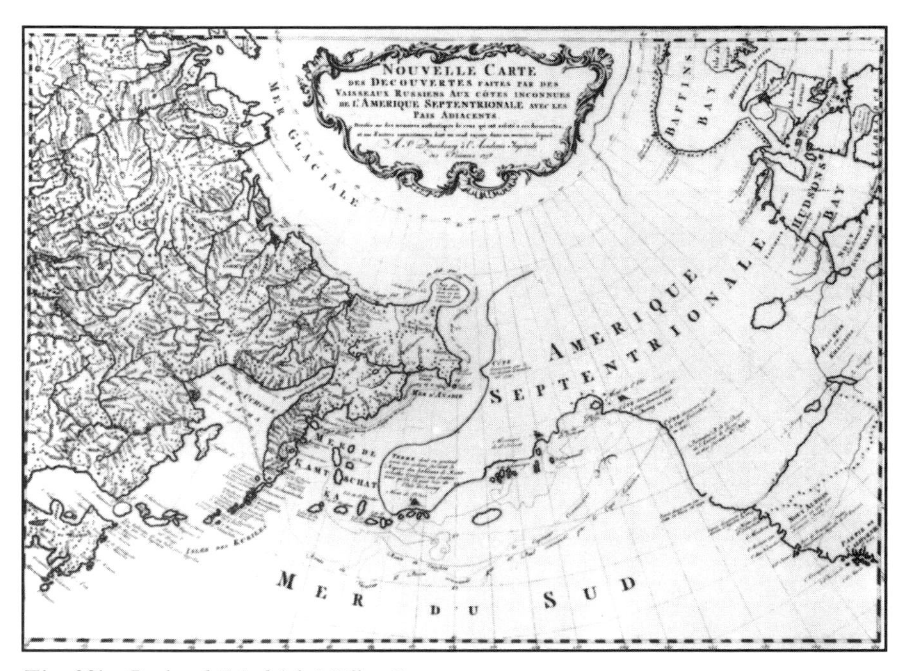

Fig. 33b. Gerhard Friedrich Müller. "NOUVELLE CARTE DES DECOUVERTES FAITES PAR DES VAISSEAUX RUSSIENS AUX CÔTES INCONNUES DE L'AMERIQUE SEPTENTRIONALE AVEC LES PAIS ADIACENTES." St. Petersburg, 1758. Copperplate, 460 x 640 mm. Private Collection.

The map presents the Russian viewpoint, showing the course of Bering's and Chirikov's voyages. The Alaskan peninsula extends far west. The apocryphal Rivière de l'Ouest runs from Lake Winnipeg to the Pacific Ocean.

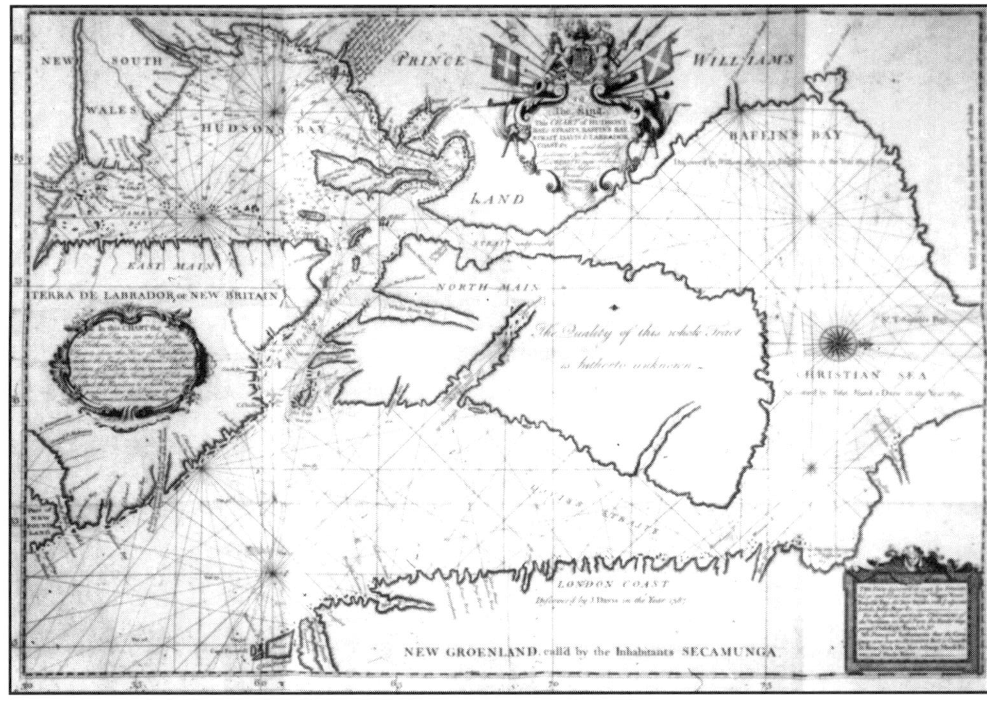

Fig. 34. Christopher Middleton. "To the King. This Chart of Hudson's Bay & Straits, Baffin's Bay, Strait Davis & Labrador Coast &c . . . ." London, 1744. Copperplate, 470 x 660 mm. Private Collection.

The map drawn as a consequence of Middleton's expedition shows that Welcome Sound did not provide an entrance into a Northwest Passage.

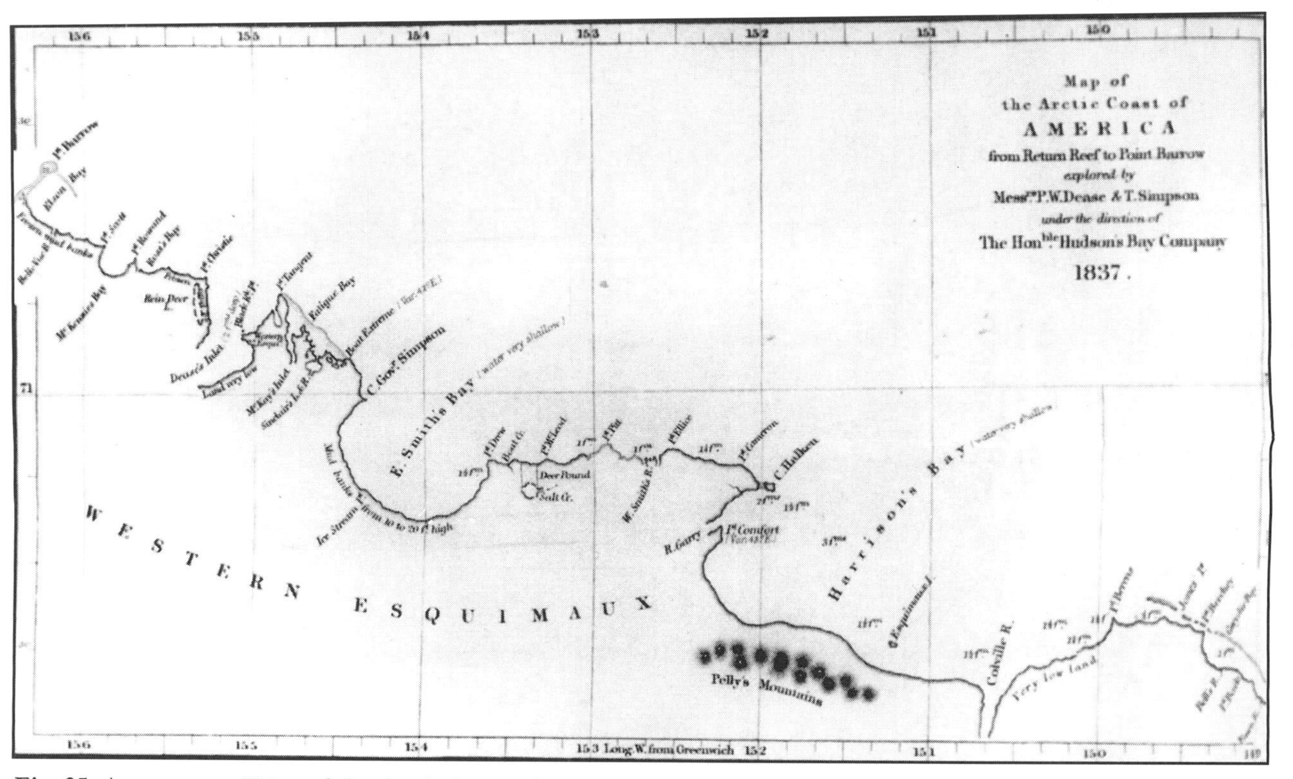

Fig. 35. Anonymous. "Map of the Arctic Coast of AMERICA from Return Reef to Point Barrow explored by Mess[rs]. P. W. Dease & T. Simpson under the direction of the Hon[ble]. Hudson's Bay Company 1837." From *Journal of the Royal Geographical Society* (London, 1838).

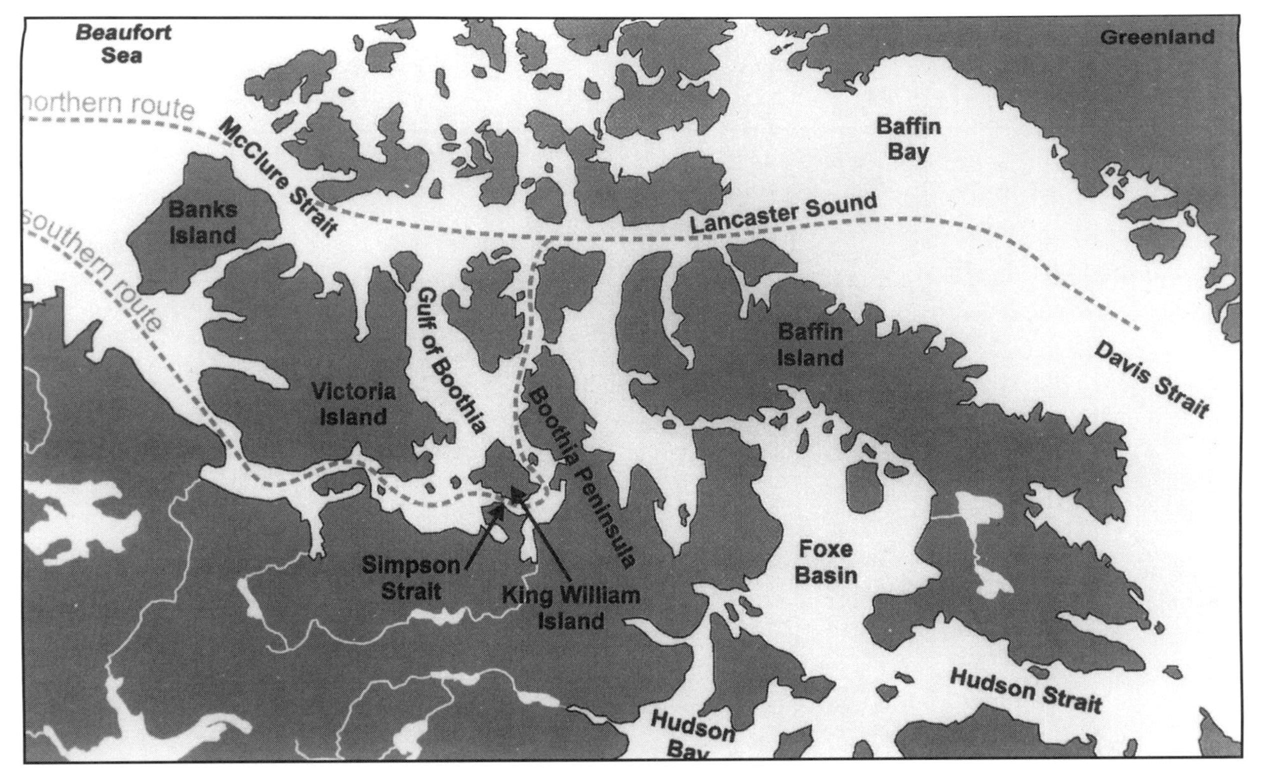

Fig. 36. "Two of the principle routes through the Canadian Arctic islands. The southern route taken by Raold Amundsen in 1903–6 was the first successfully navigated entirely by ship." From J. Victor Owen, "No Smooth Sailing: The Once and Future Northwest Passage," *Mercator's World* 7, no. 1 (2001): 30–37; reproduced with permission.

As would be expected in a tale written by a Christian man, the Californians were defeated by the Christians, and the Amazonian women became submissive. "Amazon," which derives from the Greek meaning "equal to man," was the name later given by Francisco de Orellana in 1539 to the greatest river in South America.

The etymology of the word "California" remains uncertain and continues to be argued. Did the author of *Las Sergas de Esplandian* coin the term from the Greek "kalli," meaning "beautiful," and "ornis," meaning "bird," and referring to flying griffins on the island? Or was "California" a transformation of "Califerne," a place referred to in the eleventh-century epic, *La Chanson de Roland*? Or has "California" evolved from the Arabic-Spanish "kalifon," meaning a "great province," or did the name of Caesar's wife, Caliphurnia, play a role?

If one is to ascribe the name "California" to *Las Sergas de Esplandian*, there should be adequate evidence that the tale was known to those early explorers who discovered and delineated the geography in text and on maps. There are certainly several facts attesting to the general popularity of the romantic novel, and there are strong indications that the early conquistadors who conquered Mexico and explored the west coast of North America were familiar with the book. Diego Columbus, Christopher Columbus's son, bought a copy of *Las Sergas*, which remains in the Biblioteca Columbina in Valadolid, Spain. Book 7 of the series of *Amadis* was also purchased for the library the very year it was published in Seville, indicating its general popularity.

Montalvo did not finish writing *Las Sergas* until Columbus had returned from his first voyage to the New World. One stimulus for the purchase of the book by Columbus's son could have been the writings of Peter Martyr, the humanist and tutor to the children of King Ferdinand and Queen Isabella of Spain, which popularized the reports of Christopher Columbus. In relating Columbus's experiences, Peter Martyr wrote in his journal: "Straight ahead to the north appeared a large island. Those Natives who had been brought to Spain on the first voyage, and those who had been delivered from Captivity, declared it was called Madanina, and that it was inhabited exclusively by women."

Evidence of knowledge of *Las Sergas de Esplandian* by the conquistadors in the New World can be found in the *Historia verdadera de la conquista de la Nueva España* by Bernal Díaz del Castillo. Díaz, who shared the adventures of Hernán Cortés and completed his historical tome as an octogenarian in 1572, included in his history a description of the approach by Cortés and his men to Tenochtitlán, the capital of the Aztec empire (translation by A. P. Maudslay):

> The next day, in the morning, we arrived at a broad causeway, and continued our march towards Iztapalapa, and when we saw so many cities and villages built in the water and other great towns on dry land and that straight and level causeway going towards Mexico, we were amazed and said it was like the enchantments they tell of in the legend of *Amadis*. . . .

Cortés's acceptance of the mythical island concept presented in *Las Sergas* can be found in the fourth letter he wrote to King Charles V of Spain. In the narrative of a mission made by one of his lieutenants, he wrote:

> In his descriptions of these provinces, . . . he likewise brought me an account of the chiefs of the province of Ceguatan, who affirm that there is an island inhabited only by women without any men, and that, at given times, men from the mainland visit them; if they conceive, they keep the female children to which they give birth, but the males they throw away. The island is ten days' journey from the province, and many of them went thither and saw it, and told me also that it is very rich in pearls and gold. I shall strive to ascertain the truth, and when I am able to do so, I shall make a full account to Your Majesty.

"California" appeared on a printed map for the first time in 1562 as "C. California," that is, Cape California, on the lower end of the California peninsula, of which only a small portion is shown. The map bearing the imprint is entitled "Americae sive Quarte Orbis Partis Nova et Exactissima Descriptio." It was probably drawn by Diego Gutiérrez the Younger and engraved by Hieronymus Cock in Amsterdam. Only two copies of this map exist, one in the Library of Congress in Washington, D. C., and the other in the British Library in London.

Francisco López de Gómara, in his *Historia de Mexico*, printed in 1554, introduced the word "California" in two passages concerned with Ulloa's journey. In his narrative of 1572, Bernal Díaz del Castillo used the word "California" three times: first, "It was in order to save himself the sight of so much misery that Cortés departed for the discovery of other countries and fell upon California which is a bay"; second, ". . . all the soldiers and captains whom he had left in those isles or bays which they called La California"; and third, in recounting the expedition of Ulloa, one of Cortés's lieutenants, "Cortés ordered the captain to skirt the whole extant of the coast and to sail arounc La California. . . ."

Maps made during the sixteenth through eighteenth centurie were rarely drawn by on-the-scene observers. The mapmakers genei ally relied on oral and written descriptions of the explorers, who wei not trained surveyors, and often invoked personal interpretation, imagination, or even wishful thinking. Certainly, most explorations resulted in the translation of the discoveries onto reasonably accurate charts, but, at times, errors were perpetrated and dreams delineated. Because the maps evolved from the explorers' reports, any history of the cartography of California must include a history of the expeditions that expanded the geographic knowledge of the time.

California's history stems from leadership of Hernán Cortés, one of the most famous and infamous of the Spanish conquistadors, who sequentially sailed from Cuba, founded the city of Vera Cruz, burned his ships to prevent the troops from turning back, and completed the conquest of Montezuma and the Aztec empire of Mexico in 1521. Cortés's expeditions were stimulated by the desire of King Charles V of Spain for expansion in the New World. The King's interest was increased by the description of a "great mayne sea hitherto unknowen" by Vasco Nuñez de Balboa, as viewed from his vantage point on the Darien Isthmus, now known as the Isthmus of Panama. Balboa took possession of El Mar de Sur (South Sea) in the name of the King of Spain in 1513.

Shortly after his victory, Cortés, whom a cédula (official document) of Charles V designated governor and captain-general of New Spain, was commissioned by the King to direct attention to the South Sea, so named because it bathed the south shore of the Isthmus of Panama. The goal of the search was the discovery of the legendary Strait of Anian that was referred to by Marco Polo and was thought to be the western component of a Northwest Passage. There was also hope that an expedition would uncover a land of riches and, perhaps, even the fabled island ruled by an Amazon Queen. It was logical to launch the endeavor from the western coast of Mexico, and, in preparation, a ship-building center was established in the mid-1520s at the harbor of Zacátula near the mouth of the Balsas River on the western shore of Mexico.

On July 30, 1532, the first expedition to explore northward along the coast departed Acapulco, which had become the dominant port on the west coast of Mexico. The two ships under Diego Hurtado de Mendoza failed in their mission because a mutiny occurred and the commander disappeared. A determined and tenacious Cortés dispatched two more ships under Captain Diego de Becerra the next year. This voyage was also marked by a mutiny led by the pilot, Fortún Jiménez, who killed the captain. But the mutineers continued the mission under the leadership of Jiménez and, either in late 1533 or early 1534, became the first Europeans to land on any part of Lower California, which they interpreted to be an island. The landing occurred in the region of the Bay of La Paz, just above the southern tip of the gulf shore of the peninsula. All but two of the sailors were killed by the Natives, but the two survivors, on their return to Mexico, reported their discovery of plentiful pearl beds offshore, and that there were indications of gold and precious stones in the hills, reminiscent of the mythical island in *Las Sergas de Esplandian.*

Two years after the discovery of Lower California, Cortés himself crossed the gulf and, on May 3, 1535, entered the bay later named the Bay of La Paz, where the crew from the previous mission had been killed. He took possession of the land in the name of King Charles V of Spain and established a short-lived colony there. Cortés also thought he had landed on an island, which he named Santa Cruz because he arrived on the day of the church festival of the Holy Cross. Martín de Castro, notary of the expedition, fashioned a document to formalize the act of possession, and forwarded the document to Spain, accompanied by a small sketch of the point of land and bay they had entered. That sketch was the first map of the coast of Baja California.

Shortly thereafter, Cortés, beset by political enemies in Mexico and Spain, returned to Spain. When Cortés's activities in the New World were assessed, the King made him Marqués del Valle de Oajaca and conferred upon him the title of "Captain-General of New Spain, the provinces and coasts of the South Sea, discoverer and colonizer of the coast and islands with the twelfth part of his conquests for himself and his heirs." Now armed with glory, Cortés returned to New Spain.

But before he had left for Spain, Cortés sent one of his lieutenants, Francisco de Ulloa, to sail farther north in order to find a passage around the "island" of Santa Cruz from the gulf to the South Sea, and also to search for the seven cities of Cibola, fabled for their wealth. The presence of Cibola had been described by Alvar Nuñez Cabeza de Vaca, who was shipwrecked on the shore of Galveston Bay in 1528. In the ensuing eight years, with three companions, Andrés Dorantes, Alonzo del Castillo Maldonado, and Estevanico, an Arabian black slave, de Vaca trekked across the southwestern portion of the continent, eventually arriving in Culiacan on the west coast of Mexico in May 1536. During those travels, Cabeza de Vaca performed the first chronicled operation by a European on North American soil, the removal of an arrow head from the chest of an Indian.

On July 8, 1539, Ulloa departed Acapulco in charge of three small ships and returned in seven months. The little fleet crossed into the Gulf of California, which, at the time, was called the Sea of Cortés or the Vermillion Sea. Sailing north with the east shore of the peninsula in sight, the ships reached the mouth of the Colorado, ascended the river for an undetermined distance, and took possession of the region in a symbolic ceremony. Failing to find a passage to the open sea, the fleet turned south and anchored in the harbor of Santa Cruz. After several days, they sailed north along the west coast of the peninsula, landing at Cedros Island, so named because of the cedars, and thus became the first expedition to round the tip of the peninsula. Ulloa continued north in a small vessel and reached 29° North Latitude, missing the chance of becoming the first European to view or set foot on Upper California. Ulloa's expedition was regarded as a failure because it did not discover either the Strait of Anian or the riches of Cibola. It did, however, have a major cartographic impact by suggesting that Lower California could be a peninsula. This was first depicted on a manuscript map by Battiste Agnese in 1543–44 (see figure 15).

The contemporary Spanish report of Ulloa's expedition makes no mention of the word "California." But in Ramusio's *Raccolta di Navigationi et Viaggi*, written between 1550 and 1556, the word appears three times, twice as "Isola [Island] California" in the narrative about Ulloa's voyage. Shortly after receiving Ulloa's report, the new Viceroy of New Spain, Don Antonio de Mendoza, dispatched a two-pronged exploration. Francisco Vásquez de Coronado left from Culiacan on April 20, 1540, in charge of land troops, and, on May 9, Hernando de Alarcón sailed with two ships up the gulf. His chief pilot was Domingo del Castillo, whose map is the first to show the whole outline of the Baja peninsula. Alarcón's main feat was to ascend about two hundred miles up the Colorado River. He is credited by some as being the first European to set foot on land within the current state of California. Coronado traveled overland, reaching as far as the fabled Quivira in current Kansas, but he failed to find the riches he was seeking. During the Coronado expedition, Hernando de Alvarado discovered the Rio Grande, which he named "Señora" because the river was first sighted on September 18, the traditional birthday of the Virgin Mary. Also during the expedition, a scouting party, under the command of García López de Cárdenas, was the first group of Europeans to view the Grand Canyon.

Viceroy Mendoza remained undaunted in his quest for discovering new wealth. Under his direction, on June 27, 1542, Juan Rodríguez Cabrillo sailed from the port of Navidad in Mexico with two ships, the *San Salvador* and *Victoria*. During this voyage, the coast of the current state of California was seen for the first time by Europeans. It was on September 28, 1542, after three months at sea, that Cabrillo's two ships landed at San Diego Bay, which they named San Miguel, thereby marking the discovery of Alta [Upper] California in distinction to Baja [Lower] California. Sailing north, they also landed on Santa Catalina Island, and in the region of Santa Monica they named a "Bay of Smokes." The ships rounded Point Conception above current-day Santa Barbara, and named the point "Cabo de Galera" for the prow of a galley. In the area, they sighted the islands, now known as San Miguel and Santa Rosa, and formally took possession of a port on San Miguel, appropriately naming the island, "La Isla del Posesión."

The search for the Strait of Anian continued with northerly sailing, during which neither Monterey Bay nor the Golden Gate was noted. Near current-day Fort Ross, the ships were beset with storms, turned south, and dropped anchor, on November 16, in what is now called Drake's Bay. After a short stay in Drake's Bay, the ships resumed their northern course, reentered the harbor at San Miguel, and then wintered on the islands in the Santa Barbara Channel. On January 3, 1543, Cabrillo died and was buried on San Miguel, which his companions renamed La Isla de Juan Rodríguez. The pilot, Bartolomé Ferrlo, assumed command, and, on March 1, reached the northern limit of the voyage, at approximately 41° 30' North Latitude, near the Eel River in northern California or perhaps the Rogue River in Oregon. The ships then turned south, and arrived at Navidad on April 14, 1543, almost a year after their departure. Albeit not a new term, the word "California" appears three times in the journal of the voyage, the first use of the word in a Spanish document.

These voyages provided the information and basis for the graphic representations of the western coast of North America that appeared during the first half of the sixteenth century. The first map to show a west coast of the continent was the Waldseemüller Mappemonde of 1507 (see figure 4). No representation of a Gulf of California appears on any printed map until 1544. That year Sebastian Cabot, the son of John Cabot (Giovanni Gaboto), who was the first European to set foot on North America, published a world map in Antwerp, Belgium. The only extant copy is preserved in the Bibliothèque Nationale de France in Paris. The map takes into account the explorations of Ulloa and Coronado. The upper limit of the Lower California peninsula is not delineated, leaving unanswered the question of insularity or peninsularity. Battiste Agnese's beautiful manuscript World Map of 1543–44 presents a distinct Californian peninsula and bay, but leaves the west coast of North America without a defined coast line (see figure 15).

Giacamo Gastaldi's world map (see figure 37), published in Venice in 1546, depicts in print for the first time a distinct Californian peninsula and a bay into which a long river, running north-south, empties. Gastaldi's "Carta Marina Nova Tabula," published in 1548, and his map from about 1550, "Dell'Universale . . . ," made by Matteo Pagano, show a similar geography. By contrast, the world map that appeared in Sebastian Münster's 1550 *Cosmographia* and also Antonio Salamanca's double cordiform map of about 1550, based on Gerard Mercator's 1538 map, which placed "Americae" on the North American continent for the first time, omit any representation of a Lower California peninsula.

The first map that concentrated on the New World, "Novae Insulae, XVII Nova Tabvla," (see figure 16) published by Sebastian Münster in 1540, pictures a west coast of North America without a Lower California peninsula or gulf. In Gastaldi's 1548 edition of Ptolemy's *Geographia,* the first regional map of southwest of North America depicts a Lower California peninsula and a large river emptying into the gulf, named Mar Vermio (Vermillion Sea) (figure 38).

The second half of the sixteenth century was characterized by approaches to Upper California from the west coast of New Spain. After the 1521 discovery of the Phillippines, named for King Philip of Spain, by Magellan, a Portugese sailing under the Spanish flag, most Spanish ships departed the west coast of New Spain (Mexico) for those islands and the East Indies. Acapulco became the main harbor for the Philippine trade. Richard Hakluyt, in his *Voyages,* described the port as "the harbour where the ships that goe down to China lye; and the Marchants of Mexico bring all their Spanish commodities downe to this harbour, to ship them for that countrey . . . ."

Because the Portugese controlled the route around the Cape of Good Hope, Spain had to rely on the Acapulco-Manila trade for its commercial contact. The outbound voyage from Acapulco across the Pacific was relatively easy because of favorable winds and took only two to three months. But those same winds made the return trip difficult and treacherous, extending, at times, over seven to nine months. In 1564, King Phillip II sent a Spanish fleet, led by Andrés Urdaneta and under the titular command of Miguel López de Legazpi, to determine the best route from the Orient to America. Their round trip took 129 days, and demonstrated that the best way to make a return trip to New Spain was to sail northeastward to Japan, eastward across the Pacific, taking advantage of the heavy westerly winds, at from 30° to 40° North Latitude, to the Upper California coast, and then southeastward to Acapulco. But this course required the discovery and occupation of a safe harbor on the Upper California coast, where the ships could carry out needed repairs and provisioning that followed a long transoceanic voyage.

Trading ships carried silver bullion from Acapulco to Manila, which was the collecting point for silks from the north in the Orient and spices from the south. After returning to Acapulco, many of these luxuries were transported overland and shipped to Spain from the east coast of Mexico. To preserve the market, the Spanish King limited the trade, in 1593, to one vessel a year, a 500-ton vessel that came to be known as the Manila galleon. These vessels plied the sea for over 250 years.

During the last quarter of the sixteenth century, while the Spanish were establishing the Manila trade routes, the English entered the Pacific scene. On November 15, 1577, Francis Drake left Plymouth, England in the *Golden Hind,* accompanied by four smaller vessels, with instructions from Queen Elizabeth to establish trade with the Spice Islands and also to capture treasures from the Spanish in the Indies. Drake had previously been to the New World when he sailed with his kinsman, Sir John Hawkins, to Vera Cruz in 1567, and, five years later, raided several Spanish cities in the Caribbean, at which time he crossed the Darien Isthmus in an attempt to steal silver from the royal treasure house in Panama.

True to his habit, after Drake sailed through the Strait of Magellan into the Pacific Ocean, he proceeded to plunder treasures from several Spanish ships and sacked Valpariso, Santiago, Lima, and Guatulco, Mexico. He originally planned to return to England by sailing along the coast of California, hopeful that he would find a Northwest passage through the Arctic. The ships in the fleet proceeded north along the Upper California coast, and, on June 17, 1579, anchored by the shoreline in a harbor at 38° 30' North Latitude. They remained for six days, during which Drake claimed title to the land for the Queen, and assigned the name "New Albion." The journal of the expedition indicates that the crew left behind

> a plate of brasse fast nailed to a great and firm poste, whereon is engraven her graces name and the day and year of our arrival there, and of the free giving up of the province and kingdome, both by the king and people, into her Majestie's hands; together with her highnesse picture and armes in a piece of sixpence currant English monie shewing itself by a hole made of purpose through the plate; underneath was likewise engraven the name of our general.

This so-called Drake Plate now resides in the Bancroft Library of the University of California in Berkeley, found initially in 1934 on Drake's Bay, thrown away, and rediscovered in 1936. Its authenticity has been questioned. The location of Drake's anchorage has also been the subject of debate. Some have held that it was near Point Reyes, in what is now called Drake's Bay. Others contend that the vessels anchored in Bodega Bay. After leaving the anchorage along the California coast, Drake abandoned his quest for a Northwest passage and continued his circumnavigation westward via the Moluccas, Java, and the Celebes before returning to Plymouth on September 26, 1580, almost three years after the departure. Drake went on to gain fame during the defeat of the Spanish Armada in 1588, received the title of "Sir," and died in the Caribbean Sea near Porta Bella in 1596.

Stimulated by the concern generated by the English incursion on the Pacific Ocean, the government of New Spain acted to afford added protection for their Philippine trade. In 1584, Francisco de Gali, a Manila galleon captain, was ordered to sail along the northern California coast on his return voyage from the Orient. The object was to find a port in the waters off the coast of California that could serve as a haven for replenishment. Although he was not specific, de Gali indicated that there were several safe harbors adjoining a "very fair land." At the same time, Archbishop Viceroy Pedro de Moya proposed an expedition to determine whether the North American continent was joined to Asia or if a strait called "Anian" separated the two land masses.

In the summer of 1587, Pedro de Unamuno sailed from the Philippines and, in October, landed at El Morro bay, near San Luis Obispo. There, he laid claim to the land in the name of King Philip of Spain before proceeding to Acapulco, where he arrived in late November. On July 5, 1595, a small vessel, the *San Augustín*, left Manila under the command of Sebastián Rodríguez Cermeño, who was the pilot on the Manila galleon *Santa Ana* when it was captured by Thomas Cavendish in 1587. The *San Augustín* made landfall at Cape Mendocino, a little north of current Eureka, California, on November 4. The cape was named to honor either L. Suárez de Mendoza, Viceroy of New Spain from 1580 to 1583, or Antonio de Mendoza, Viceroy in 1542, the year it was supposedly discovered. Cermeño's ship then entered the same bay in which Drake had anchored sixteen years before, and Cermeño took possession of the region in the name of the King of Spain. The ship was later driven ashore by a squall and destroyed. The undaunted crew built a launch in which they completed their voyage, arriving in Navidad on January 7, 1596.

The second half of the sixteenth century witnessed dramatic

changes in the mapping of California, based on the reports emanating from the explorations that were completed. The woodcut map (see figure 39) accompanying *Primera y segunda parte de la Historia general de las Indias . . .* by Francisco López de Gómara, published in Zaragosa, Spain, in 1553, was the first Spanish printed map to depict the west coast of the North American continent. This rare map shows a remarkably accurate west coast, including a Lower California peninsula and gulf. Adjacent to the lower tip is "C. de Vallenas," a misspelled Ballenas, meaning "Cape of Whales."

In 1554, Michele Tramezzino published a copperplate map in Italy presenting a gulf running parallel to the Pacific coast and ending in a point at 37° North Latitude where a north-south river enters. A similar representation appears on the Giacomo Gastaldi–Gerard de Jode map, published in Antwerp in 1555, and on maps by Jeronimo de Girava, Caspar Vopell, and Paolo Forlani, all published during the same decade. Gastaldi's world map of about 1561 improved the accuracy of the delineation of the west coast. It includes a Lower California peninsula and bay, and is the first map to show an unnamed strait separating Asia and America. By contrast, Forlani's 1562 and 1565 maps maintain a broad connection between the two large continents. A large world map, published in Antwerp in 1564, was the first cartographic production of Abraham Ortelius, who would lead the Low Countries to supremacy in the field of cartography. That map followed a delineation that had been introduced on Giacomo Gastaldi's 1561 world map.

Doubtless the most recognizable name in the field of cartography during the Renaissance was that of Gerard Mercator, whose great world map of 1569 introduced the projection that bears his name. Mercator's projection presents the surface of the globe on a flat plane. By increasing the distance between the parallels of latitude towards the poles, the correct relationship between angles on the map results, and loxodromes are represented as straight lines. This allowed seamen to lay down a compass course as a straight line. Shortly before he died in 1594, Mercator proposed bringing out an edition to be entitled *Atlas sive Cosmographicae Meditationes de Fabrica Mundi et Fabricati Figura.* The title incorporated the first use of the term "Atlas" for a collection of maps. Perhaps the term was adopted from the title page of a compilation of maps with varied formats by Lafreri that first showed, in 1560, the symbolic figure of the Titan Atlas supporting the world on his shoulders. Mercator's 1569 world map depicts a Californian coast extending from the lower peninsula, with its adjacent gulf, to the Strait of Anian, which leads to a wide Northwest Passage.

In 1570, for the first time, a modern Atlas with a uniform format was published. Abraham Ortelius of Antwerp compiled fifty-three maps of the world in *Theatrum Orbis Terrarum.* The world map, "Typus Orbis Terrarum," and the map of the Western Hemisphere, "Americae Sive Novi Orbis, Novus Descrptio" (see figure 23), both offer a reasonable west coast of North America, a Lower California Peninsula, and an adjacent gulf, into which a large river, extending from the mountain chain in the middle of the continent, drains. This graphic representation persists essentially unchanged on world maps for the remainder of the sixteenth century. Joannes Myritius's woodcut map of 1590 stands out as a rare exception in showing Asia and North America connected by a broad land mass.

In general, maps of the Western Hemisphere that were published during the second half of the sixteenth century depict the same geographic representation of California that has been described for world maps. But these more focused maps introduced unique elements and additions. The *Theatrum Orbis Terrarum: Terzo Volume delle Navigationi et Viaggi* by Giovanni Battista Ramusio was a widely read and influential book. It includes "Universale Della Parte Del Mondo Nvovamente Ritrovata," the first printed map of Amerca to include any of the names from the 1540–42 travels of Francisco Vásquez de Coronado. A California peninsula is notable, but the northwestern portion of the continent is blank, leaving the question of a land bridge or strait between Asia and America unanswered. As indicated previously, the Diego Gutiérrez 1562 work, where the word "California" first appeared on a printed map, presents only the tip of the peninsula, and none of the west coast of the peninsula or continent..

In 1566, Paolo Forlani of Venice published the first map specifically focusing on the North American continent (see figure 22). "Labrador," "New France," "Apalchen," "Florida," "Civola," and "Quivira" all appear in large print. The seven cities of Civola (Cibola) was a locale in the region of current New Mexico, which, the Native Americans told the conquistadors, was a rich source of gold. In 1539, Fray Marcos reported to Mendoza, the Viceroy of New Spain, that they were merely seven Zuni pueblos with the sun reflecting on them, and no gold was present. The "Sierra Nevada," meaning snowy mountains, are also named. With the exception of the Gastaldi maps, the Anian Strait is shown for the first time on the Forlani map, which, in addition, assigns the name to the strait, adopting a term introduced by Marco Polo and referred to in a 1562 pamphlet by Giacomo Gastaldi. The map also depicts Lower California peninsula and gulf, into which a major river complex discharges.

Michael Lok's woodcut map, which played an important role in the cartographic history of the False Sea of Verrazzano and the Northwest Passage (see figure 18), presents a long and broad Californian peninsula with a narrow isthmus separating the gulf from the open seas leading to the Orient. The map mimics the 1554 delineation drawn by Michaelo Tramezini, and it is probably one of progenitors of the "California as an Island" concept that became accepted forty-three years later.

In 1597, Cornelis van Wytfliet of Louvain published *Descriptionis Ptolemaicae Augmentum,* a compilation of nineteen maps concerned only with the New World. In effect, this is the first Atlas of the Americas. Among the eight maps relating to North America, there appears "Granata Nova et California," the first printed map devoted to California and the southwest of what is currently the United States of America (see figure 40). Included on the map are the seven cities (of Civola), "Mar Vermeio" (Vermillion Sea), "Californiæ Sinus," "C. de California," and "Y. de Cedros," named during the Ulloa voyage. The word "California" extends from the main land mass into the peninsula around which the map is centered. The coast of Upper California is presented with an erroneous westward slant. The other map of the west coast included in atlas is entitled "Limes Occidentis Quiura et Anian 1597," and covers the entire west coast of Canada and the United States, extending from an open sea in the Arctic Circle southward to 30° North Latitude. "C. de Mendocino" is the most western point of the continent. Two "C. Blanco"s are named, as is a "C. de S. Francisco." (see figure 41).

In 1600, Gabriel Tatton, a noted hydrographer of London, had two elegant copper engravings published in Amsterdam by Cornelis Claez. The map of the southwest, "Noua et Rece Terrarum et Regnorum Californiæ Nouæ Hispaniæ Mexicanæ, et Peruviæ . . ." differs from its companion map by including an imaginary lake with a river flowing out of it into the Gulf of California. "California" appears in bold print on the peninsula. The other Tatton map, "MARIS PACIFICI" (see figure 42), presents the west coast of North America with a long westerly slant. An extension to the Anian Strait is accompanied by a legend that reads "De este Cabo Mendocino Hasta el Estrecho de Anian esta por Descubrir," indicating that the coast is yet to be discovered. Both "California" on the peninsula and "Nova Albion" in the northwestern part of the continent are noted. Henry R. Wagner, the authority on the Spanish voyages and the mapping of the western part of North America, has suggested that the names appearing on the west coast of the Tatton maps previously had been incorporated on the Emery Molyneux globe of 1592, while other names were derived from the voyages of Juan Cabrillo.

After an interlude of almost half a century, during which ships ceased sailing north from New Spain in an effort to better define the west coast of Lower and Upper California, the seventeenth century opened with renewed resolve. In 1602, the Viceroy of New Spain, the Conde de Monterey, dispatched an expedition to survey and map the west coast of the continent as far north as Capo Blanco at 44° North Latitude. The Viceroy's instructions specifically indicated that the gulf was not the object of exploration, that permanent settlements should not be established, and that the names that had been assigned previously should not be changed. On May 5, 1602, the expedition led by Sebastián Vizcaíno and consisting of two large ships, the *San Diego* and the *Santa Tomás*, and the frigate *Tres Reyes*, departed Acapulco. Gerónimo Martin Palacios, a cosmographer, was aboard to chart the coast in detail. Three friars participated in the mission; one of these, a Camelite Fray Antonio Ascensión, was also a cosmographer.

After reaching the Bay of San Bernabé at Cape San Lucas at the tip of the peninsula, the ships continued north, and, on November 10, anchored in the harbor that Cabrillo had named San Miguel. Disregarding the Viceroy's specific instructions, Vizcaíno renamed the port "San Diego" to honor the saint on whose day the ships arrived. San Diego was the fifteenth-century Spanish friar to whom King Philip was devoted, and was also the name of Vizcaíno's flagship. The excellence of the port of San Diego was reaffirmed, and the chronicler of the voyage asserted that the fish and game were plentiful and that there was evidence of gold on the beach. The small fleet sailed north to an island, which they named Santa Catalina, to honor St. Catherine, patron of Christian philosophers and the saint on whose day the discovery was made. The fleet anchored in a cove in the region of the current Avalon resort, and then proceeded north through the Santa Barbara channel on December 4, naming that channel for the patron saint of artillerymen.

On December 8, the day of the Immaculate Conception, the crew viewed a headland that Cabrillo had named Galley Cape because it resembled a galley's prow. Vizcaíno replaced that name with "Conception." On December 12, "Point Reyes" (Kings' Point) was named for the Three Kings who offered their adoration to the Child Jesus on Twelfth Night.

The ships next anchored in a port that they considered optimal for replenishment of the Manila galleons. Vizcaíno suggested that a settlement should be established to assist those ships, and pointed out that an added attraction was that it was of a climate and quality similar to Spain. That same harbor, which had been called the "Bay of Pines" by Cabrillo and the "Bay of San Pedro" by Cermeño, was renamed "Monterey" by Vizcaíno to honor the Viceroy. The descriptions of the area by both Vizcaíno and Ascensión were so full of exaggerations that the Portola expedition was unable to recognize the bay when they viewed it in 1789. One of Vizcaíno's vessels, the *Santa Tomás*, was sent back to announce the discovery of the harbor, and the remaining two ships, the *San Diego* and the *Tres Reyes*, continued north, becoming separated.

The *San Diego* anchored in Drake's Bay, passed Cape Medocino, and turned south for home because of the poor condition of the crew, who were dying from starvation and scurvy. Vizcaíno, accompanied by five men, went ashore at Mazatlán, where they were given agave, a small fruit that miraculously improved the symptoms of scurvy in a few days. The *Tres Reyes* reached farther north to Cape Blanco, where the boatswain, in a later statement, indicated the coast ran northeast and led to the Strait of Anian, through which passage to the Atlantic Ocean could occur, leading Fray Ascensión to the conclusion that the Gulf of California joined the Strait to the north, thereby making California an island. He wrote:

> In this part this realm has north of it the Kingdom of Anian, and to the east the land which is continuous with the realm of Quivira. Between these two realms extends the strait of Anian, which runs to the North Sea, having joined the Oceanic Sea which surrounds Cape Mendicino and the Mediterranean Sea of California, both of which are united at the entrance of the strait which I call Anian. Toward the west is the realm of China, and toward the south all the realm of Japan.

The captain of the *Tres Reyes* died, and only six men were alive when the small vessel reached Navidad. The expedition's voyage lasted eleven months and succeeded in providing detailed descriptions and maps of the coast as far north as Monterey. The transition to a new Viceroy, the Marqués de Montesclaros, dramatically altered interest in the exploration or colonization of California. From the time of Vizcaíno's voyage in 1602–3 until 1789, no ship sailed the waters off the coast of California from New Spain under the auspices of the government.

About the same time, Don Juan de Oñate of Zacatecas, whose wife was the granddaughter of Cortés and the great-granddaughter of Montezuma, marched north from Mexico to the Rio Grande. In 1598, the group left Chihuahua to establish a colony in New Mexico. During a series of expeditions, Oñate and his companions passed near current El Paso and established the regional capital, called San Juan de los Caballeros, north of current Bernalillo. They reached the pueblo at Acoma, the oldest place of continuous habitation in the United States, and proceeded into the Quivira region as far as Wichita, Kansas. Inscribed on a rock in western New Mexico are the words that translate: "Passed by here the Governor Don Juan Oñate to the discovery of the Sea of the South on the sixteenth of April the year 1606." Oñate was replaced as governor of the region by Pedro de Péralta, who moved the capital in 1610 to Santa Fe. By 1620, Santa Fe had fifty residents and was second only to St. Augustine as a locale in the current United States permanently inhabited by Europeans and, subsequently, by Americans. During their journey, they reached the mouth of the Colorado River and the gulf. They were informed by the Natives that there was a body of water beyond the mountains that blocked their view. They interpreted this to be an extension of the gulf farther north. The Natives also reported the presence of an Amazon queen and a land replete with gold, silver, and pearls.

Throughout most of the first quarter of the seventeenth century, essentially all world and Western Hemisphere maps depicted a Lower California peninsula and a west coast with a northwesterly slant. World maps by the leading cartographers, such as Jodocus Hondius, Peter Plancius, Willem J. Blaeu, Pieter van den Keere, and Claes Janszoon Visscher—all working in Amsterdam, the dominant center of carotography—presented a consistent representation of the North American continent and its western portion. The consistency was also true for the maps concentrating on the Western Hemisphere.

The first departure from this representation and the first time California appeared as an island was on the title page of Michael Colijn's 1622 edition of *Descriptio Indiæ Occidentalis* (see figure 43 [showing the 1623 Frankfurt edition]). The book is a translation of the original work of Antonio de Herrera y Tordesillas that was published in 1601. As Henry R. Wagner asserted, the small vignette, measuring 9 x 12 cm, can hardly be construed as showing an island of California. Rather, there is a long river running into the Gulf of California from the north, and that river is prolonged to connect with an open sea along the northwest coast of America. The basis for the map is unknown, but Wagner ascribes it to a manuscript map attached to a document by Nicolas de Cardona, who traveled to the upper part of the gulf and the mouth of the Colorado River at 34° North Latitude. "Descripcion De Las Yndias Occidentalis," the larger map of the Western Hemisphere that appears in *Descriptio Indiæ Occidentalis* shows the previously prevalent contemporary representation of California as a peninsula.

Chronological accuracy based on the date of publication must credit "'t Noorder deel van WEST-INDIEN" by Abraham Goos as the first major map to depict California as a distinct island (see figure 44). The 19 x 29 cm copperplate engraving appears in *West-Indische Spieghel* by Athanasius Inga, considered to be a pseudonym of an unknown author. The work was published in Amsterdam in 1624, and the text includes no reference to California. The map is the first to name the "Hudson River," "De la war bay" for an early Virginia Governor, and "Cape Codd." Wagner and other authorities strongly assert that the Goos map is a derivative of the Briggs's map of 1625, even though it has an earlier publication date. The terms newly introduced on the Goos map are of English origin, and are all present on the Briggs map with the same spelling. Both maps include evidence of the 1602–3 expedition of Sebastián Vizcaíno, including "P. S. Diego," "S. Clement," and "S. Catalina," all for the first time.

The map by Henry Briggs (see figure 45) first appeared in Purchas's *Pilgrimes*, published in London in 1625. The map is generally regarded to be the progenitor of the most persistent cartographic misrepresentation relating to the North American continent, California as an island. In the lower left corner of the map, a legend states:

> California sometimes supposed to be a part of ye westerne continent but since by a Spanish Charte taken by ye Hollanders it is found to be a goodly Islande: the length of the west shoare being about 500 leagues from Cape Mendocino to the South Cape thereof called Cape St. Lucas: as appeareth both by that Spanish Chart and by the relation of Francis Gaule whereas in the ordinaire Charts it is sett downe to be 1700 Leagues.

The Briggs map is the first printed chart to include the new nomenclature and discoveries of the Vizcaíno expedition of 1602 and 1603 up the west coast of California. A "R. del Norte," representing the Rio Grande is shown emptying into the Gulf of California instead of the Gulf of Mexico. This is evidence of the cartographers' lack of knowledge of the findings of Oñate, who had shown that the Rio Grande ran east or southeast rather than west. The "Rio del Norte," or Rio Grande, was first correctly delineated as flowing into the Gulf of Mexico on a map in Vincenzo Coronelli's *Atlante Veneto* of 1690. On the Briggs map, near Point Reyes, a port bears the name "Puerto Sir Francisco Draco." This seems to be an attempt to offer the

Spanish equivalent of Drake's name. The location is identical with that named "Puerto San Francisco" by Sebastian Cermeño in 1595.

Wagner concluded that the concepts related to California, as depicted on the Briggs map, emanated from the views expressed by Fray Antonio Ascensión, who accompanied Sebastián Vizcaíno. Ascención indicated in the manuscript that he sent to Spain in 1620 documenting the Vizcaíno expedition that it was accompanied by a map, but the map has not been found. Ascención's text, however, declares that the Gulf of California extends to the north to a connection with the Strait of Anian.

Wagner's assertion that the Goos map derived from Briggs's concepts and an earlier map by Briggs, which has not been found, is based on a piece written in 1622. Attached as an appendix to the *Declaration of the State of the Colony and Affairs in Virginia*, there appeared "A Treatise of the North-West Passage in the South Sea, through the Continent of Virginia and by Fretum Hudson." The work was signed "H. B." In support of the concept that the Pacific was but a short distance from Virginia, Briggs stated that Button's Bay "doth extend itselfe very neere as farre towards the west as the Cape of *California*, which is now found to be an Iland stretching it selfe from 22 degrees to 42 degrees and lying almost directly North & South; as may appeare in a map of that Iland which I haue seene here in *London*. . . ." It seems likely that Briggs sent a copy of his map to a publisher in Amsterdam in 1622, and Goos adopted Briggs's interpretation of California as an island.

Henry Briggs's treatise was probably written as a consequence of his role as a member of the Trading Company to England. Briggs's "official biography" in the first edition of Great Britain's definitive *Dictionary of National Biography* makes no mention of his only, albeit critical, cartographic contribution. Doubtless, his reputation as an intellect was a major factor in acceptance of the revolutionary map. Henry Briggs had received a Bachelor and Master of Arts from Cambridge. His first post was that of Professor of Geometry at the newly established Gresham College in London, a post that he held from 1596 to 1619, when he assumed the Sevilian Professorship at Merton College, Oxford. He was a most distinguished mathematician, who published *Arithmetica Logarethmecia* in 1624, and his *Trigonometria Britannica* was published posthumously in 1633.

Joannes de Laet, in his *Novus Orbis, seu Descriptionis Indiae Occidentalis*, Libri XVIII, published in 1633, states that he had seen an old manuscript map showing California as an island in an Atlas. He discounted the theory and included a map showing a Californian peninsula. But most seventeenth-century maps depict California as an island, and, in time, the concept became more pervasive. Among maps of the world published between 1625 and 1650, John Speed's first English world map of 1626 shows California as an island. By contrast, Jean Boisseau of Paris published some maps with a California island and others on which the coastline is left open, avoiding the issue. Still other cartographic firms, such as those of Hondius in London and Visscher in Amsterdam, offered some publications depicting California as an island and others as a peninsula. On maps published during the second half of the seventeenth century, California is shown as island more frequently, and, during the last quarter of that century, essentially all major world maps include an island of California.

An increasing representation of California as an island also pertains to maps focusing on North America. In 1650, a landmark map of the continent by Nicolas Sanson, who initiated the rise in French cartography, includes a distinct island, "Californie Isle," extending from "C. Blanc" to "C. de S. Lucas" with a "Mar Vermejo" separating the island from the continent.

Maps showing California as an island have been the subject of an ever-expanding interest of cartographic historians. The first focused checklist of these maps was produced in 1964 by Ronald Vere Tooley, who included one hundred examples, exclusive of world maps. This seminal work, which was the most popular topic published in the Map Collector's Series, was expanded by John Leighly in 1972, in part by including world maps, resulting in 182 separate maps. Finally, in 1995, *The Mapping of California As an Island* by Glen McLaughlin included 249 printed maps (excluding world maps), seventeen title pages and frontispieces, and six celestial charts containing maps, all showing California as an island.

As the seventeenth century drew to a close, a large island of California was firmly imprinted not only on maps but also in the minds of those with an interest in the area. At the same time, a man of heroic proportions, whose efforts would figuratively reattach the island to the North American continent, entered the scene. That man, who incorporated religious zeal, a knowledge of mathematics and cartography, an adventuresome spirit, and an inquiring mind, was Father Eusebio Francisco Kino (see figure 46).

Father Kino, who played an integral role is establishing the truth about California's geographic status, was born in Segno, in northern Italy on August 10, 1645. He studied at the Jesuit College in Trent, Italy, and continued his studies at the Jesuit College in Hall, Austria. Kino entered the Jesuit Order on November 20, 1665, and, five years later, finished his course of philosophy at Inglostadt, Austria. Two of his professors at Inglostadt, Adam Aigenler, a cartographer and mathematician, and Heinrich Scherer, a cartographer and geographer, contributed significantly in his education and participated in the promulgation of his discoveries and maps. At the time he completed his studies, Kino wrote to the Jesuit General, requesting a foreign mission. But before his request was granted, Kino taught literature for three years at Hall, Austria, and returned to Inglostadt for four years of theological studies while continuing his education in mathematics and the natural sciences.

Father Kino was ordained in 1677, and, after a year of training in Bavaria, he finally left for his foreign mission. Following a long series of delays in Spain, he sailed across the ocean, arriving at Vera Cruz on May 1, 1681. Ironically, Father Kino arrived in Mexico City that year believing that California was a peninsula and not an island. But his subsequent review of the reports of Don Juan de Oñate and the contemporary maps by European cartographers caused Kino to temporarily change his opinion. He would, however, over the course of time, prove beyond a reasonable doubt that California was *not* an island.

Kino's maps were integral elements in his reports detailing his many explorations. The maps complimented or substituted for written text. Over a period that extended from 1683 through 1703, Father Kino produced the most numerous and significant series of maps in the history of New Spain. Kino was so wedded to graphic expression that he even drew maps to enforce his preachings to his Native subjects. His scientific training led him to define accurately the latitude of the areas he was charting. Chronologically, Kino's maps became increasingly more accurate and inclusive, correcting previously perceived facts and eliminating errors.

On January 17, 1683, Admiral Atondo y Antillón, with Father Kino as one of two Jesuits included in the party, left from Ciacala, Mexico, by ship, and crossed the Gulf of California to the bay of Nuestra Señora de Guadalupe, where they built a short-lived settlement. Later that year, Kino joined a second expedition that set out for the mouth of the Rio Grande in Lower California, and built a fort and mission at San Bruno. After exploring the interior of the peninsula, they established a small agricultural colony. Two of Kino's maps related to these expeditions are preserved in the Archives of the Indies in Seville. One is a map of part of Lower California, entitled "Delineacion de la Nueva Provincia de S. Andres, del Puertode la Paz, y de las Islas circumuecinas de las Californias o Carolinas . . ."; the other is a plan of Fort Bruno.

On the map of Lower California, the Gulf of California is named

"Mar de las Californias ò, Carolinas," honoring the reigning King Charles II of Spain. Only a small segment of Lower California is depicted. The two provinces named are "S. Andrés" and "SS[a] Trinidad." To the west of the landmass that is designated "Parte de las Californias, ò Carolinas" is a small part of the Pacific Ocean, called "Mar del sur."

In December 1684, Kino was part of an expedition that crossed the Lower California peninsula, and viewed the Pacific Ocean. In January, the explorers christened a river emptying into the ocean "Santo Tomás," honoring the Mexican Viceroy. Once again, Kino amplified his written description with a map, which he sent Father Heinrich Scherer, his Professor at Inglostadt College. Scherer redrew it, and incorporated the map in his monumental *Atlas Novus*, published in Augsburg in 1702. The map centers on the gulf, named "Mar Vermeio o de las Californias," and depicts the lower portion of the peninsula and the opposite Mexican western provinces. The peninsula extends only as far north as the recently named "Río de St. Thoma." Lower California is designated as "Pars Insulae Californiae" (part of the California Islands). A translation of the Atlas's Latin text reads:

> California is the first and principal island of the entire world and also the largest. It is separated from New Mexico by the narrow Red Sea—termed "Vermejo" in Spanish. Lengthwise from south to north it spans about 24 degrees latitude: that is 360 leagues, which are equivalent to 480 French leagues or 432 Spanish leagues: converted into miles, 1,440 Italian or English miles.

Plans for expanding settlement and the establishment of more missions along the eastern shoreline of the Gulf of California were temporarily interrupted by the financial demands of Spain's war with France. At the end of 1686, Father Kino set out from Mexico City on a four-month trek to the north, with the goal of establishing his own mission to be used as a base for expansion. He carried with him a painting, "Nuestra Señora del los Dolores," by the Mexican artist Juan Correa. In the vicinity of Bamotze in the valley of the San Miguel River, Kino established the mission that would become his home base and would serve as the nucleus for his efforts at expansion from March 13, 1687, to the time of his death on March 15, 1711. Kino died in the Sonoran chapel of Magdalena, where he had gone to officiate at its dedication. The mission of Dolores, which was located about one hundred miles from Tucson, was christened with the name of the painting that Father Kino had transported from Mexico City.

During the eight-year period between the establishment of his base mission and the year 1695, Kino conducted fourteen expeditions and established many Native settlements, converting and providing assistance to the Natives. In 1690, he began his series of explorations to the north. In 1692, he reached the San Pedro River, and, on December 15, 1693, he arrived at Caborca on the east shore of the Gulf of California, which he looked across and assigned names to the mountains that came into his view: San Marcos, San Mateo, San Juan, and San Antonio. A year later, in the company of Juan Mateo Manje, he reached the Gila River and rediscovered the Casa Grande ruins, which had been previously found by Francisco de Ibarra in 1597. Father Kino descended the river for a distance of four leagues, where he came upon a Native village on November 30, and named it San Andres, the patron saint of the day. San Andres or St. Andrews was one of the Apostles, who was martyred at Patriae in about 60 A.D. He remains the patron saint of Scotland, and the St. Andrew's cross is distinctive in that it is shaped like the letter X.

After a rebellion of the Pima Indians in 1695, during which one of his fellow missionaries was killed, Father Kino returned to Mexico City to plead for the continuance of his efforts on behalf of the Natives. While in Mexico City, Kino completed two maps to be incorporated as illustrations for the biography of Father Francisco Javier Saeta, who lost his life in the Pima Indian revolt.

The first of the two maps in the biography is entitled "Theater of the Apostolic Efforts of the Society of Jesus in North America" (see figure 47), and is preserved in the Central Jesuit Archives in Rome. California is depicted in the fashion of Fray Antonio de la Ascensión and Henry Briggs, as a large island, extending from "C^O^ Blanco," above "C^O^ Mendozino" south to "C^O^ de S. Lucas." The island bears the names "Californias" and "Carolinas," while the gulf is named "Mar de Las Californias." The map, however, focuses on the northern part of western New Spain, in the region west of the Gulf of California. This heavily annotated segment extends from "Guadalaxara" to Río del Tizon and includes Acoma (pueblo) and numerous missions and settlements.

In the lower left corner there is a legend that lists a series of expeditions related to California, including the following (translated):

> Don Fernando Cortés set out in 1533 as the first to discover California, landing at the port of Nuestra Señora de la Paz. . . . General Francisco de Alarcón . . . General Sebastián Biscaíno went there in 1596. . . . In 1606 he received a royal decree order-

> ing him to found a settlement in the port of Monterey. In 1615 Captain Juan Iturbi made two expeditions to California. Captain Francisco Ortega over to California the first time in 1632 . . . Captain Carboneli undertook his expedition a few years after Captain Ortega. . . . Luis Cestín de Cañas crossed over in 1642. . . . Admiral Don Pedro Porter Casanate disembarked in 1644 in the bay of San Bernabé. . . . In 1664 Admiral Don Bernardo Bernal de Piñadero made an expedition in two ships . . . and in 1672 he received orders to make second expedition, which he completed. In 1668 Captain Francisco Lucenilla made an expedition in two ships. . . . Admiral Don Isidro de Atondo y Antillón, in 1681, 1682, 1683, 1684 and 1685 . . . crossed over to the Californias and even to the opposite shore . . . last year (1694), when Captain Francisco Itamarra went to the Californias, the Natives asked most insistently for the Fathers of the Society of Jesus. On December 19, 1693, we beheld from this land of the Pimas and coast of New Spain the nearby region of California; and again on November 27, 1694, at the thirty-fourth parallel, we discovered the pleasant and productive Río Grande del Coral which pours its vast volume of water into the arm of the Sea of California. . . .

The second map in the biography of Father Francisco Javier Saeta resides in the Central Jesuit Archives, and depicts the martyrdom of Father Saeta, kneeling at an icon of "S. Diego del Pitquín and Concepción de Caborca" about to receive the arrows of two Indians. The mouth of the Rio Grande del Coral as it discharges into the gulf is shown and the names of several Native tribes and many missions are included.

The ultimate seed of the concept that California was a peninsula and not an island was planted in Father Kino's mind during a trip he made in February, 1699. On that occasion, he descended the Gila River, and viewed its junction with the Colorado (a name applied by the Oñate party in 1604 and meaning "color-red") River. Kino noted Natives with abalone shells, which he had previously seen on the Pacific coastal shore, and he concluded that the shells could only have been brought across over a land passage. In September 1700, he viewed the entrance of the Colorado River into the gulf, and thereby satisfied himself that California was not an island. In March 1701, in order to provide additional proof, Kino set out to follow the western shore of the gulf to the mouth of the Colorado, but was unable to complete the mission.

Kino's conclusion that California was not an island was correct but its proof was not complete. Even his associate and fellow explorer, Juan Mateo Manje, wrote in his diary that Kino had not unequivocally established the peninsularity of California. To define that a peninsula existed, it would be necessary to complete a journey along the east coast of the gulf to its end at the mouth of the Colorado River, or to traverse the land between the Colorado River and the Pacific Ocean. In 1746, Father Fernando Consag provided the final proof when he reached the mouth of the Colorado River by a journey along the coast of the Gulf of California.

Father Kino's most famous and influential map was drawn in 1701, and provided a graphic representation of his conclusion that is expressed in the map's title: "Passage by land to California." The map was the result of nine explorations made between 1698 and 1701, which had as their main objective the determination of whether California was an island or a peninsula. The quest grew out of the need to find a land route by which supplies could be transported to newly initiated missions on Lower California from the established missions on the mainland, and also the desire to extend the missions into Upper California.

No original manuscript map known to have been drawn in either 1701 or 1702 has been found, but it is known that Kino sent a 1701 manuscript map to Father Bartolomé Alcázar, professor of mathematics at Colegio Imperial in Madrid. Alcázar redrew that map in preparation for printing, and sent it to the highly regarded French engraver, Charles Inselin, who was commissioned by the French Jesuits to make the plate for printing. The map first appeared in print in 1705 in the popular *Lettres édifiantes,* and, a few months later, in the more scientific *Mémoires de Trévoux* (see figure 48). A printed version of Kino's 1702 manuscript map, which includes a few relatively minor additions to the 1701 map, was first published in the 1726 edition of the German mission magazine, *Der Neue Welt-Bott.*

The two French printings adhere more closely to a literal translation of the Spanish manuscript: "Passage by Land to California, discovered by the Reverend Jesuit Father Eusebio Francisco Kino from 1698 to 1701, where one can see the new Missions of the Fathers of the Society of Jesus." The German printing of the "Map of California in the year 1702, drawn by Reverend Father Kino, S. J., in accordance with his personal observations," adds a chauvinistic subtitle "Land Passage to California, discovered and delineated by Reverend Father Eusebio Francisco Kino of the Society of Jesus, a German; indicated are the new Missions of the Society, from the year 1698 to 1701." The latitudes incorporated on the three maps are 25°30' to 36°30'. The previously included alternative for "California," namely "Carolinas" is absent for the first time, following the death of King Carlos II and the accession of Philip V.

Father Kino's monumental and revolutionary 1701 map presented several cartographic innovations. Most importantly, it laid to rest the previously pervasive concept of California's insularity. Kino established California as a peninsula, based on his personal experiences; the declaration, in this instance, was not based on hearsay evidence. The location, course, and confluence of the Gila and Colorado Rivers were more precisely defined, as were some of their tributaries. The map also presented the newly established Jesuit missions on the Peninsula for the first time, and provided the encouraging information that there was a relatively short distance between those missions and the central established missions in northwestern Mexico.

The printed versions of the 1701 and 1702 maps were widely distributed and transformed the cartography of California. But before considering this revolutionary transformation, mention should be made of the influence of Kino's earlier maps. Kino's 1683 map of the area around the bay of La Paz, at the southern end of the Baja peninsula, was incorporated into the maps of many European mapmakers. Although Claude Delisle, the French King's official geographer, regarded as inconclusive Kino's map of 1695–96 depicting California as a large island, this did not prevent Nicolas de Fer, geographer of his Highness the Dauphin, from publishing a map in 1700 that pirated the Kino 1695–96 map, without mentioning the source, in his *L'Atlas curieux*.

The map by de Fer is entitled "Cette Carte De Californie et Du Nouveau Mexique . . . Avec privilege du Roy 1700" and includes 314 numbered locations on the mainland accompanied by a key to the numbers. In 1720, de Fer published another edition of the map with an altered title: "La Californie ou Nouvelle Caroline. Teatro de Los Trabajos Apostolicos de La Compa. E Jesus en la America Sept^{e}," on which the place names appear on the map proper.

The expression on a widely disseminated map, emanating from a distinguished and authoritative cartographer, was apt to be adopted. In France, de Fer's model was followed by Jacques, Chiquet, and later by the influential Guillaume de l'Isle, son of Claude, in his "Hemisphere Septentrional." That map does incorporate elements from Kino's 1705 printed map. English maps produced by Herman Moll, John Senex, George Willdey, Richard Mount & Thomas Page, Edward Wells, Henry Overton, George Foster, Edmund Halley, and Thomas Kitchin, all depict California as an island, following the representation proposed by Henry Briggs. Some European cartographers continued to show California as an island as late as 1784. Japanese maps perpetuated this presentation into the nineteenth century. A map by Shuzo Sato, published in 1865, depicts California as a large island, oriented nearly east-west.

The 1716 edition of Patrick Gordon's *Geography Anatomiz'd: Or the Geographical Grammar*, published in London, included in its descriptions the following narrative:

> *California*
>
> This Island was formerly esteem'd a *Peninsula*, but now found to be intirely surrounded with Water. Its North Part was discovered by Sir *Francis Drake*, Anno 1577, and by him call'd *New Albion*, where erecting a Pillar, he fasten'd thereon the Arms of *England*. The Island parts thereof were afterwards search'd into, and found to be only a dry, barren, cold Country, *Europeans* were discourag'd from sending Colonies to the same, so that it still remains in the Hands of the Natives; and there being nothing remarkable relating either to them or on it, we shall proceed to, Newfound-Land.

Father Eusebio Francisco Kino's rediscovery of the peninsularity of California, and the expression of that concept on a distinct map, established his reputation. In 1707, a German edition of Kino's map was printed in Leipzig and Frankfort. In 1708, the map appeared in *The Philosophical Transactions for the Months of November and December*. Each of the five London editions (1794 through 1808) of *A General Atlas* by Thomas Kitchin includes a redrawing of Kino's 1701 map. The inset on the map reads:

> A new map of North America. The passage by land to California, discover'd by Father Eusebio Francis Kino, a Jesuit; between the years 1698 and 1701, before which, and for a considerable since, California has always been described in all charts & maps as an island.

The map with the same legend was repeated in four editions (1768 through 1782) by T. Jefferys in his *The American Atlas*.

But the concept of California as an island was hard to erase, as evidenced by its persistence on European maps until late in the eighteenth century. This presentation persisted despite a formal royal decree by Ferdinand VI of Spain that "California is not an island." The history of the mapping of California and its passage through several different geographic renditions is encapsulated on a map by Robert de

Vaugondy published in 1779 in *Receuil de 10 cartes,* and also in the *Supplément* to Diderot's popular *Encyclopédie* (see figure 49).

After Kino drew his most famous map, he established seven new missions between 1703 and 1707. In 1706, Kino chronicled his discovery of the islands of Santa Ynes and San Vicente in the Gulf of California. Manuel de Oyuela, a Franciscan, recorded viewing a large island in the mouth of the Colorado River while in the company of Kino. From 1708 until his death, Kino designated northern Mexico "Nueva Navarra" to indicate that the region linked New Spain with New France, in the same way that Navarre linked Spain and France in Europe.

These addenda have been noted on a map discovered by Ernest J. Burrus, S. J., in the middle of the twentieth century. In sorting through the D'Anville collection of maps in the Bibliothèque Nationale de France in Paris, Burrus uncovered a manuscript that he regarded to be a redrawing of a map made by Father Kino during his later years. The title of the map, translated from its original Spanish, is "New Kingdom of the New Navarre and the contiguous other kingdoms, 1710." In the upper right corner an inscription reads: "Copied from the original manuscript, September 9, 1724." Lower California is named "Penisla de California," and contains the legend: "Newly ascertained to be such by Father Kino of the Society of Jesus in the discovery which he made in 1702." The innovations that have been noted on the newly discovered manuscript found expression on many printed maps, including those by Guillaume de l'Isle, Jean-Baptiste Bourguignon D'Anville, Charles-Marie Rigobert Bonne, Robert de Vaugondy, Johann B. Homann, Alexander von Humboldt, Tomás López, and Thomas Kitchin..

Father Eusebio Francisco Kino's name is perpetuated by "Kino Crater" in the Santa Clara Mountains; "Kino Bay" on the Sonoran mainland; "Punta Kino," the northern point of Sierra San Nicólas; "Cerro Kino" in the Lower California peninsula; and "Kino Peak," the highest point in Pima County, Arizona.

The three swings of a pendulum, from island to peninsula, from peninsula to island, and finally from island to peninsula, are history. Will the distant future add a fourth swing, transforming California into an island? About thirty million years ago, a shift of the tectonic plates in the earth's crust began. The movements resulted in fracture of the crust, called a "fault." The notorious fault, known as the San Andreas Fault, ironically bears a name that is a variant of San Andres, the name given to a Native settlement in 1693 by Father Eusebio Kino, the man who proved that California was not an island.

The San Andreas Fault (see figure 50) is the dominant member of a network of faults, known as a "fault system," that extends over a distance of more than six hundred miles, from Point Arena, west of San Francisco Bay, to the Gulf of California. The Hayward, Calaveras, San Jacinto and Elsinore Faults are included in the "system." The memorable San Francisco earthquake of 1906 and the subsequent lesser quakes in the same region are episodes of the displacement that has been about 350 miles over the years since the first shift occurred. The Pacific plate, on the west, has slipped, and, at times, lurched northwestward relative to the North American plate on the east.

California currently remains a part of the contiguous forty-eight United States of America. The royal decree issued in 1747 by King Ferdinand VI of Spain that "California is not an island" stands. Will the westward slippage of the western tectonic plate eventually create an island of California?

References

Anghiera, Pietro Martire d'. *De orbe novo: The Eight Decades of Peter Martyr D'Anghera.* Translated by F. A. MacNutt. 3 vols. New York: Putnam, 1912.

Ascensión, Antonio. *Monarchia Indiana.* Seville, 1615.

Ascensión, R. P. Fray de la. "Relación de la jornada que hizo el Sevastien Vizcayno del descubrimiento de las Casifornias el año de 1620." In Henry R. Wagner, *Spanish Voyages to the Northwest Coast of America in the Nineteenth Century.* San Francisco: California Historical Society, 1929.

B., H. [Briggs, Henry?] "A Treatise of the North-West Passage in the South Sea, through the Continent of Virginia and by Fretum Hudson." In *Declaration of the State of the Colony and Affairs in Virginia.* London, 1622.

Briggs, Henry. *Arithmetica Logarethmecia.* London, 1624.

———. *Trigonometria Britannica.* Goudae, 1633.

Coronelli, Vincenzo. *Atlante Veneto.* Venice, 1690.

Cortés, Hernán. *The Five Letters.* Translated by F. A. MacNutt. New York: Putnam, 1908.

Diáz del Castillo, Bernal. *Historia verdadera de la conquista de la Nueva España.* Madrid, 1632. Translated as *The True History of The Conquest of New Spain.* Translated by A. P. Maudsley. London: The Hakluyt Society, 1908–16.

———. *The True History of the Conquest of New Spain.* Translated by A. P. Maudslay. 5 vols. London: The Hakluyt Society, 1908–16.

Fer, Nicolas de. *L'Atlas curieux.* Paris, 1700–1703.

Gordon, Patrick. *Geography Anatomiz'd: Or the Geographical Grammar.* London,

1716.

Hakluyt, Richard. *Divers Voyages Touching the Discovery of America*. London, 1582.

Hale, Edward Everett. "The Queen of California." *Atlantic Monthly* vol. XII, no. LXXVII. Boston: 1824, 265-78.

Herrera y Tordesillas, Antonio de. *Novus Orbis, sive Descriptio Indiae Occidentalis*. Amsterdam, 1622. Translated into German as *Zwölfter Theil der newen Welt, das ist: Gründliche volkommene Entdeckung aller der west indianischen Landschafften, Insuln*, . . . Frankfurt, 1623.

Inga, Athanasius. *West-Indische Spieghel*. Amsterdam, 1624.

Jefferys, T. *The American Atlas*. 4 editions, 1768–82.

Kino, Eusebio. *Vide del P. Francisco Javier Saeta, S. J.* Mexico City: Editorial Jus, 1961.

Kitchin, Thomas. *A General Atlas*. 5 editions. London, 1794–1808.

Laet, Joannes de. *Novus Orbis, seu Descriptionis Indiae Occidentalis*, Libri XVIII. 1633.

Lobeira, Joâo de. *Las Sergas de Esplandian*. In *Amadis de Gaule*, Book 5. London, 1664.

López de Gómara, Francisco. *Historia de Mexico, con el descubriemento de la Nueva España*. Antwerp, 1554.

———. *Primera y segunda parte de la Historia general de las Indias* . . . . Zaragosa, 1553.

McLaughlin, Glen. *The Mapping of California As an Island: An Illustrated Checklist*. California Map Society Occasional Paper No. 5. Saratoga, Calif.: California Map Society, 1995.

Mémoires de Trévoux: Mémoires pour l'histoire des sciences et des beaux arts. Paris, 1701–67.

Mercator, Gerard. *Atlas sive Cosmographicae Meditationes de Fabrica Mundi et Fabricati Figura*. Duisberg, 1595, and many later editions.

Münster, Sebastian. *Cosmographia*. Basel, 1550. *Der neue Welt-Bott* (1726).

Philosophical Transactions of the Royal Society of London for the Months of November and December 1708.

Piccolo, Francisco, and Charles Le Gobien. *Lettres édifiantes*, écrites des missions étrangères par quelques missionaires de la Compagnie de Jésus. Paris, 1705.

Ptolemy. *Geographia*. Edited by Gastaldi. Venice, 1548.

Purchas, Samuel. *Purchas His Pilgrimes*. London, 1625.

Ramusio, Giovanni Battista. *Navigationi et Viaggi*. Venice, 1556–88.

———. *Theatrum Orbis Terrarum: Terzo Volume delle Navigationi et Viaggi*. Venice, 1570.

Robert de Vaugondy, Gilles. *Recueil de 10 cartes*. Livorno, 1779.

Scherer, Heinrich. *Atlas novus*. Augsburg, 1702.

Southey, Robert. *Amadis of Gaul*. 4 vols. London: T. N. Longman and O. Rees, 1803. From the Spanish by Carciordonez de Montalva.

Tooley, Ronald Vere. *California As an Island: A Geographical Misconception*

Illustrated by 100 Examples from 1625 to 1770. Map Collectors' Series, no. 8. London: Map Collectors' Circle, 1964. Expanded by John Leighly in 1972: *California as an Island.* San Francisco: Book Club of California.

Wagner, Henry R. "Father Antonio de las Ascencion's Account of the Voyage of Sebastian Vizcaino." In his *Spanish Voyages to the Northwest Coast of America in the Nineteenth Century.* San Francisco: California Historical Society, 1929.

Wytfliet, Cornelis van. *Descriptionis Ptolemaicae Augmentum.* Louvain, 1597.

Selected Readings

Burrus, Ernest J., S. J. *Kino and the Cartography of Northwestern New Spain.* Tucson: Arizona Pioneers' Historical Society, 1965.

Cleland, Robert Glass. *From Wilderness to Empire: A History of California, 1542–1900.* New York: Alfred A. Knopf, 1944.

Davidson, George. "The Origin and Meaning of the Name of California." *Transactions and Proceedings of the Geographical Society of the Pacific* Series 2, vol. 6, part 1 (1910): 3–50.

Lutgens, Frederick K., and Edward J. Tarbuck. *Essentials of Geology.* 6th edition. Upper Saddle River, N.J.: Prentice-Hall, 1998.

McLaughlin, Glen. *The Mapping of California As an Island: An Illustrated Checklist.* California Map Society Occasional Paper No. 5. Saratoga, Calif.: California Map Society, 1995.

Putnam, Ruth. *California: The Name.* University of California Publications in History, Vol. 4, No. 4. Berkeley: University of California Press, 1917.

Rolle, Andrew F. *California: A History.* New York: Thomas Y. Crowell Company, 1963.

Wagner, Henry R. *The Cartography of the Northwest Coast of America to the Year 1800.* 2 vols. Berkeley: University of California Press, 1937.

———. "Some Imaginary California Geography." *Proceedings of the American Antiquarian Society* (April 1926).

———. *Spanish Voyages to the Northwest Coast of America in the Nineteenth Century.* San Francisco: California Historical Society, 1929.

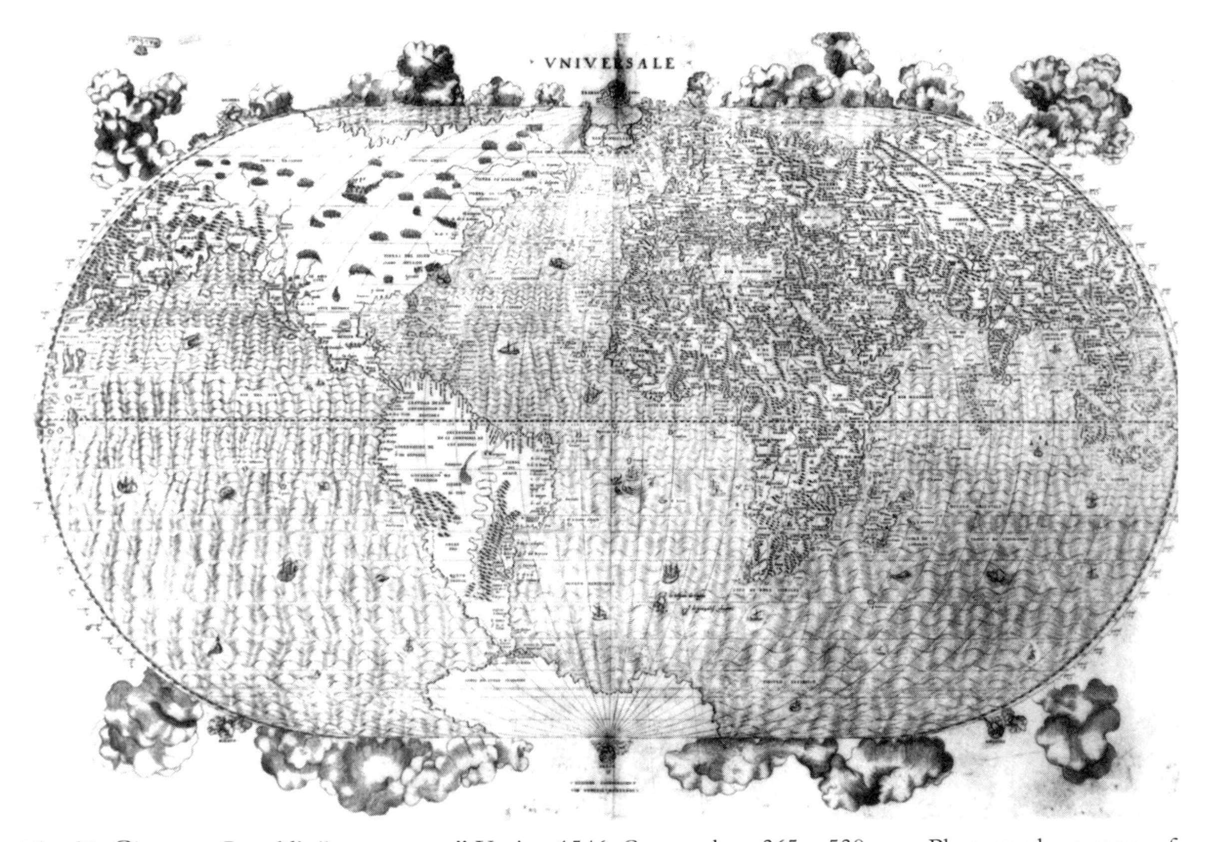

Fig. 37. Giacomo Gastaldi. "VNIVERSALE." Venice, 1546. Copperplate, 365 x 530 mm. Photograph courtesy of John Carter Brown Library, Brown University, Providence, R.I. The first map to depict a California peninsula.

Fig. 38. Giacomo Gastaldi. "NVEVA HISPANIA TABVLA NOVA." Copperplate, 130 x 175 mm. From Ptolemy, *La Geografia* (Venice, 1548). Private Collection.

The first regional map of the southwest of North America. A large river, "R. tontonteanc" (representing the Colorado or Gila River) empties into the "Mar Vermeio" (Vermillion Sea).

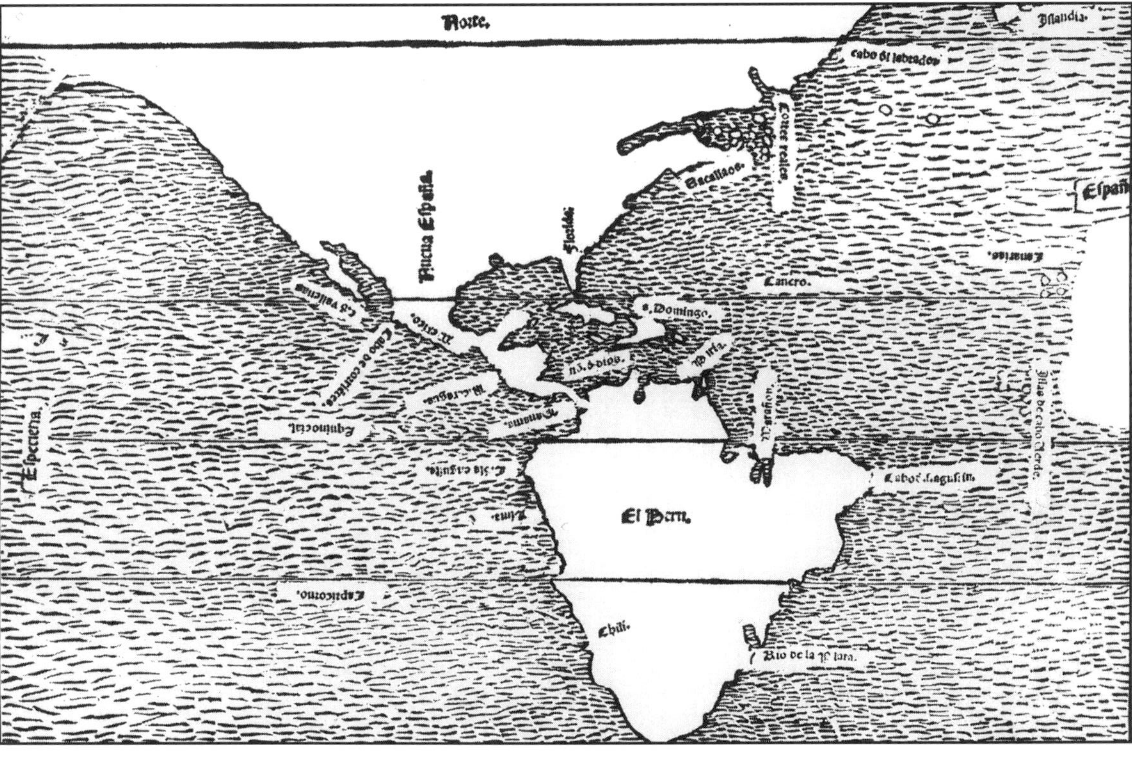

Fig. 39. Francisco Lopez de Gómara. [No Title]. Woodcut, 190 x 290 mm. From *Primera y segunde parte de le la historia general de las Indias* . . . (Zaragoza, Spain, 1552). Private Collection.

The first Spanish map to depict the entire North American continent. "C. de Vallenas," a misspelling of Ballenas, refers to the Cape of whales.

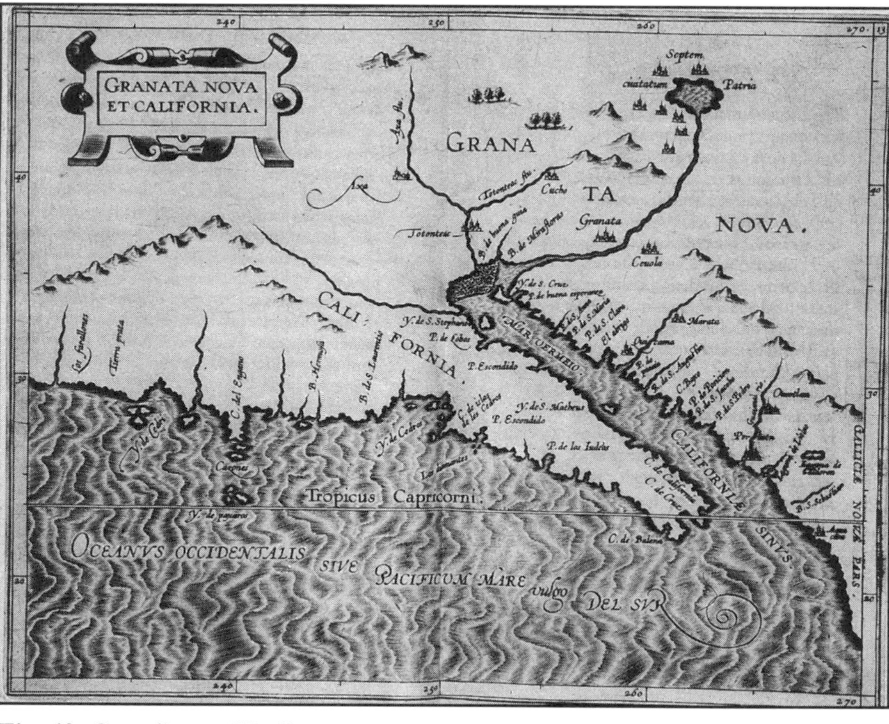

Fig. 40. Cornelis van Wytfliet. "GRANATA NOVA ET CALIFORNIA." Copperplate, 230 x 290 mm. From Wytfliet, *Descriptionis Ptolemaicae augmentum* (Louvain, 1597). Private Collection.

The earliest printed map devoted to California and the southwest. "California" extends from main continental landmass into the lower peninsula. "Californie Sinus" and "C. de California" are named.

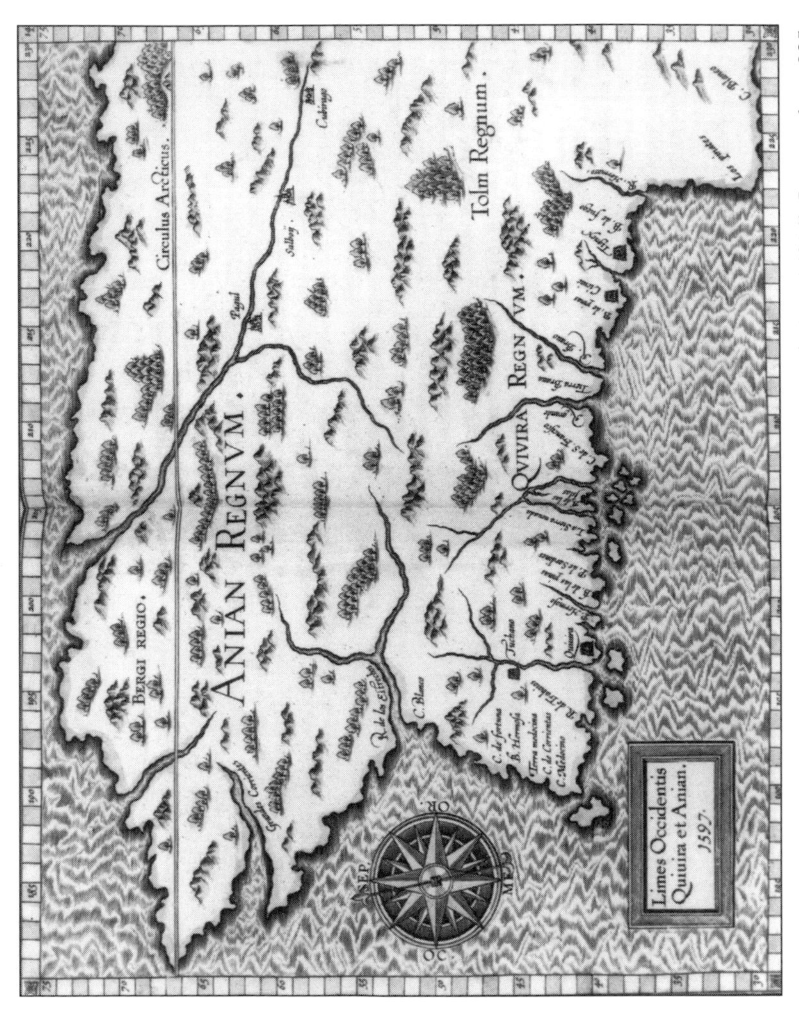

Fig. 41. Cornelis van Wytfliet. "Limes Occidentis Quiuira et Anian. 1597." Copperplate, 235 x 290 mm. From Wytfliet, *Descriptionis Ptolemaicae augmentum* (Louvain, 1597). Private Collection. The map contains the names "C. Mendocino," two "C. Blanco," and a "C. de S. Francisco."

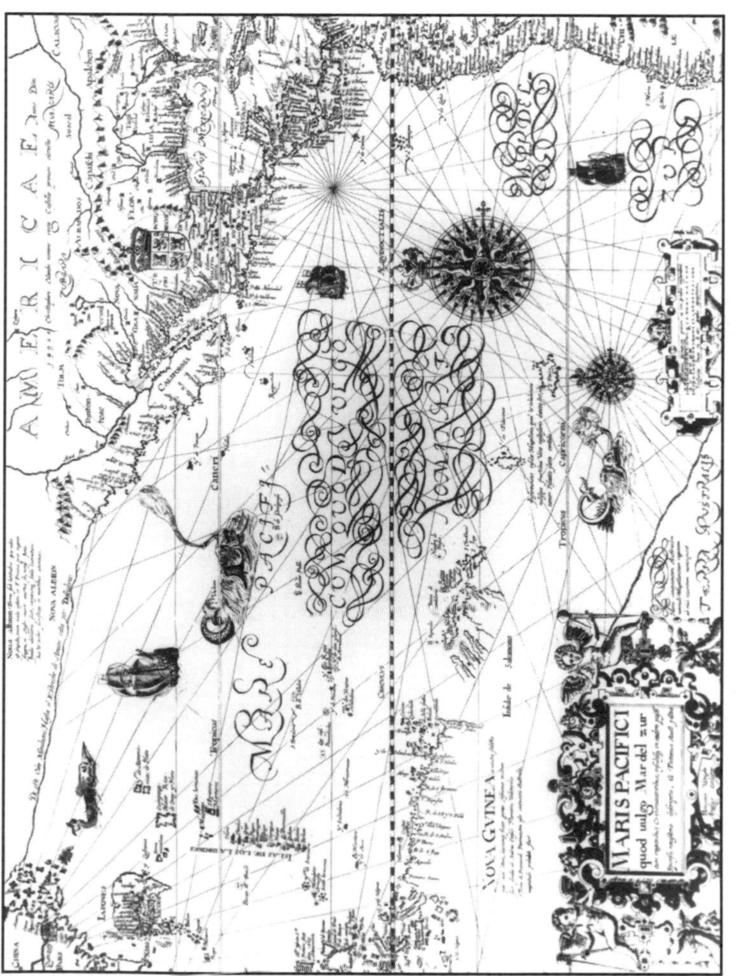

Fig. 42. Gabriel Tatton. "MARIS PACIFICI . . ." Amsterdam, 1600. Copperplate, 405 x 530 mm. Private Collection.

The west coast of North America has a long westerly inclination, and the legend indicates that the coast is yet to be discovered.

Zwölffter Theil
der Newen Welt/
Das ist:
Gründliche volkommene Entdeckung
aller der West Jndianischen Landschafften/ Jn-
suln vnd Königreichen/ Seecusten/ fliessenden vnd stehenden
Wassern/ Port vnd Anlendungen/ Gebürgen/ Grentzẽ/ vnd
Außtheilung der Provincẽ/ sampt eygentlicher Beschreibung
der Stätte/ Flecken vnd Dörffer/ Herrschafft vnd Regirung/
Bistummen/ Stifft vnd Clöster/ wie starck dieselben an Jn-
wohnern/ wie reich an Einkommen/ was jedes Orts Gewerb/
Handthierung vnd Bequemlichkeiten/ Fruchtbarkeit vnd Nu-
tzung/ alles nach jetziger Gestalt vnd Beschaffenheit von
newem entdecket vnd beschrieben/
Durch
Antonium de Herrera, Königlichen bestellten Histo-
rienschreiber der Reiche Castilien vnnd Jndien/ auß der
Hispanischen Sprach in die Teutsche vbergesetzet.
Jtem
Gewisse Anzeig der jenigen/ so durch die gefährliche Enge der
Magellanischen Strassen oder Sunds hindurch passirt/ vnd
den Erdt Kreiß rings vmbfahren haben.
Jtem
Petri Ordonnez de Cevallos Beschreibung der West Jndia-
nischen Landschafften/ sampt andern Anhängen.
Alles mit schönen Landtaffeln vnd Kupfferstücken vor Au-
gen gestellt zu volkommener Erklärung der obbe-
sagten Materien.
Gedruckt zu Franckfurt/ in Verlegung Johann Dietherichs
de Bry/ Anno 1623.

Fig. 43. Michiel Colijn. [No Title]. 1622. Copperplate, 90 x 95 mm. From Antonio de Herrera y Tordesillas, *Zwölfter Theil der newen Welt, das ist: Gründliche volkommene Entdeckung aller der west indianischen Landschafften, Insuln* . . . (Frankfurt, 1623). Private Collection.

The small map on the title page of the book is the first to suggest that California was an island.

Fig. 44. Abraham Goos. " 't Noorder deel van WEST-INDIEN." Copperplate, 185 x 285 mm. From Athanasius Inga, *West-Indische Spieghel* (Amsterdam, 1624). Private Collection.

The first major map to depict California as an island. It is generally believed that the author drew on the Briggs map (figure 45) as its source. The map is also the earliest one to name "Hudsons River," De la war bay," and "Cape Codd." The map includes "P. San Diego," "S. Clement," and "S. Catalino," all named during the Vizcaino expedition of 1602–3.

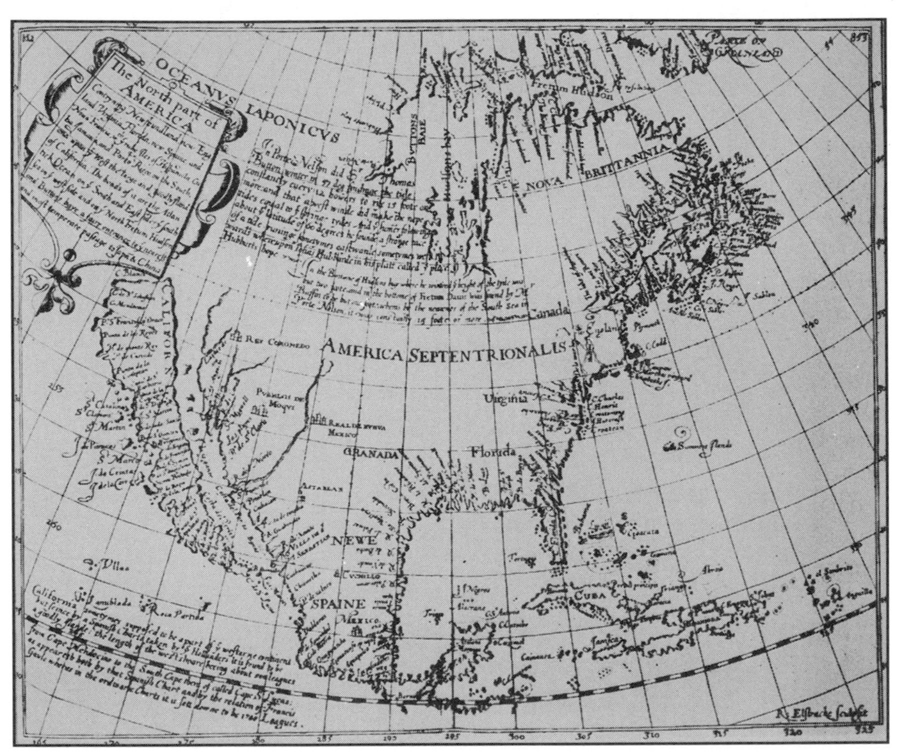

Fig. 45. Henry Briggs. "The North part of AMERICA Conteyning Newfoundland, new England, Virginia, Florida, new Spaine, and Noua Francia, wth y^{e} riche Iles of Hispaniola, Cuba, Jamaica, and Port Rieco, on the South, and upon ye West, the large and goodly Iland of California . . ." Copperplate, 285 x 350 mm. From Samuel Purchas, *Purchas His Pilgrimes* (London, 1625). Private Collection.

This map drawn by a distinguished mathematician accompanied his "Treatise of the North-West Passage to the South Sea, through the Continent of Virginia, and by Fretum Hudson." The map is considered to be the true progenitor of the concept of California as an Island.

Fig. 46. Father Eusebio Francisco Kino (1645–1711). Bronze by Suzanne Silvercruys. Placed by the State of Arizona in National Statuary Hall, photograph by Karsh, Ottawa, 1965.

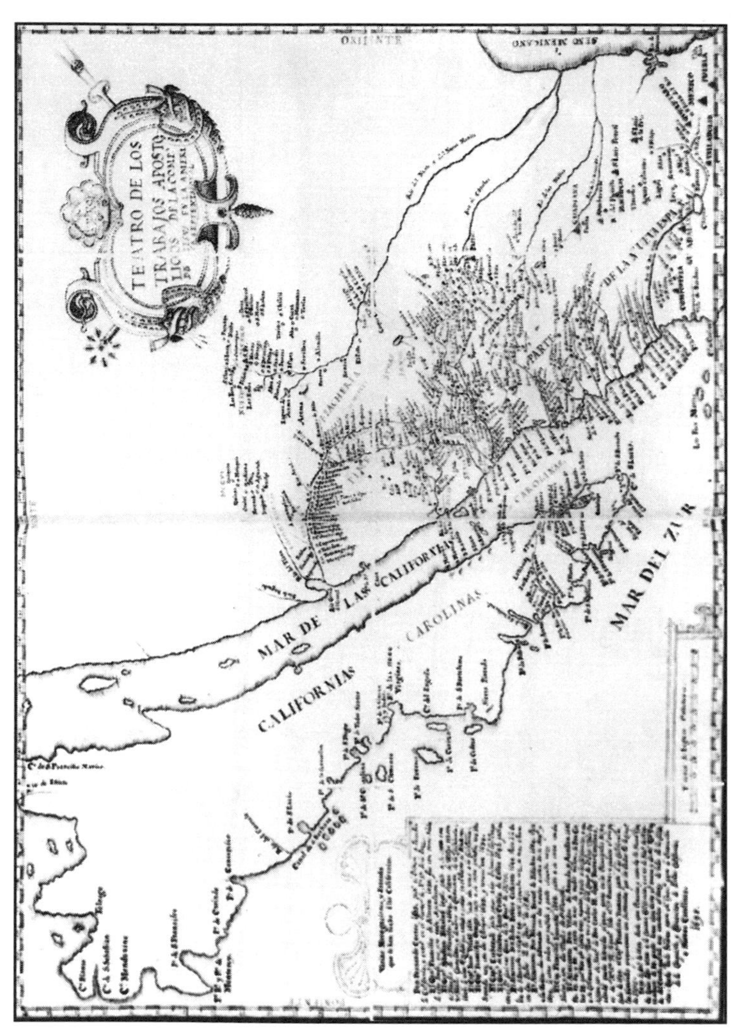

Fig. 47. Father Eusebio Francisco Kino. "Theater of the Apostolic Efforts of the Society of Jesus in North America, 1695" [translated]. Manuscript, 825 x 600 mm. Photograph courtesy of Central Jesuit Archives, Rome.

The map depicts Kino's early revised concept that California was a large island.

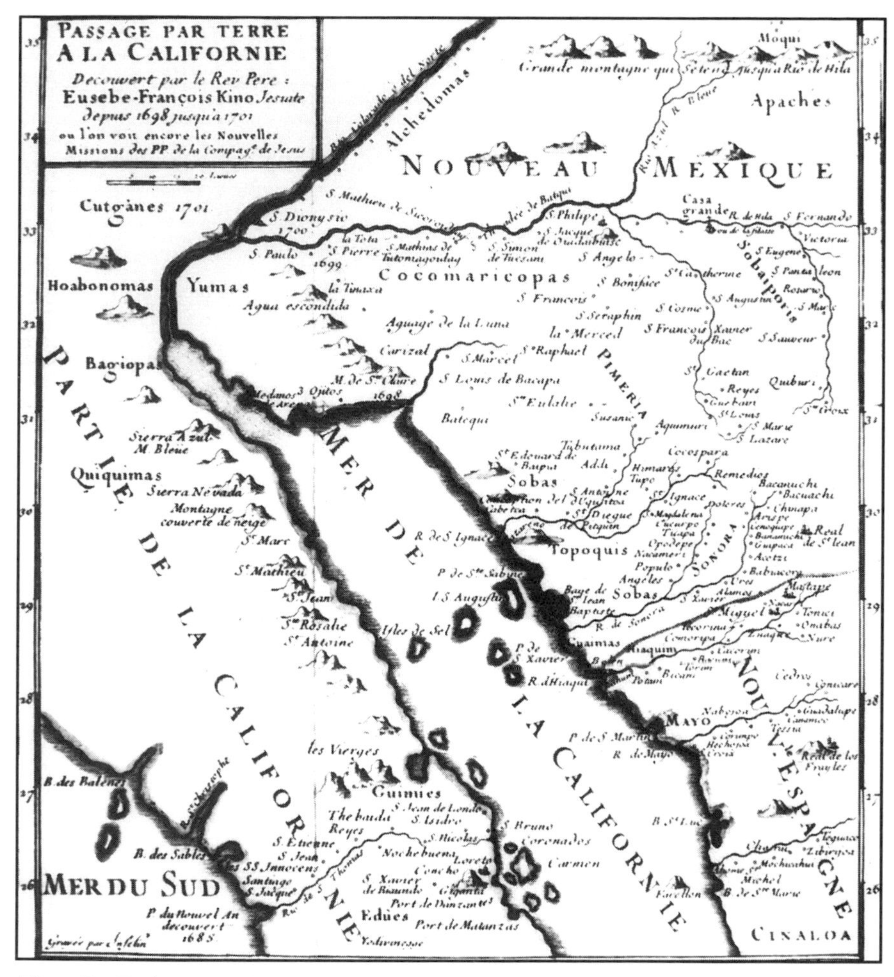

Fig. 48. Father Eusebio Francisco Kino. "PASSAGE PAR TERRE A LA CALIFORNIE Decouvert par le Rev. Pere Eusebe-François Kino Jesuite depuis 1698 jusqu'a 1701 . . ." Copperplate, 240 x 210 mm. From *Mémoires de Trévoux* (Paris, 1705). Private Collection.

The print was made by Charles Inselin from a 1701 manuscript map sent by Kino to Father Bartolomé Alcázar in Madrid. It represents Kino's conclusion, based on personal explorations, that California was connected to the mainland.

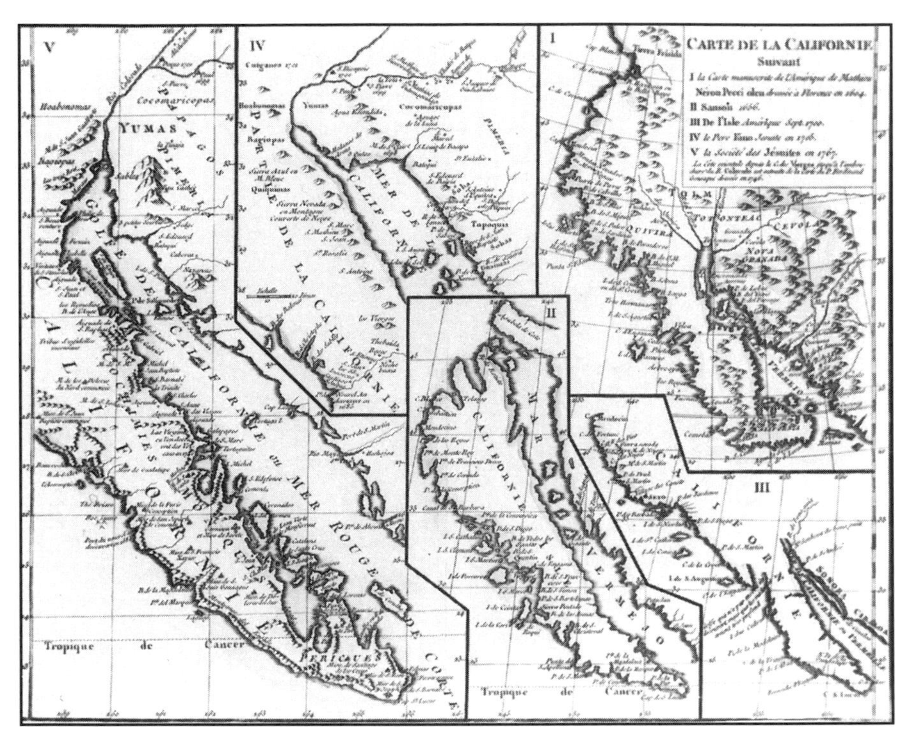

Fig. 49. Robert de Vaugondy: "CARTE DE LA CALIFORNIE . . ." Copperplate, 290 x 370 mm. Paris, 1779. From Robert deVaugondy, *Recueil de 10 cartes traitant particulièrement de l'Amérique du Nord* (Livorno, 1779). Private Collection.

The map traces the history of mapping of California with representative maps from 1604, 1656, 1700, 1705, and 1767.

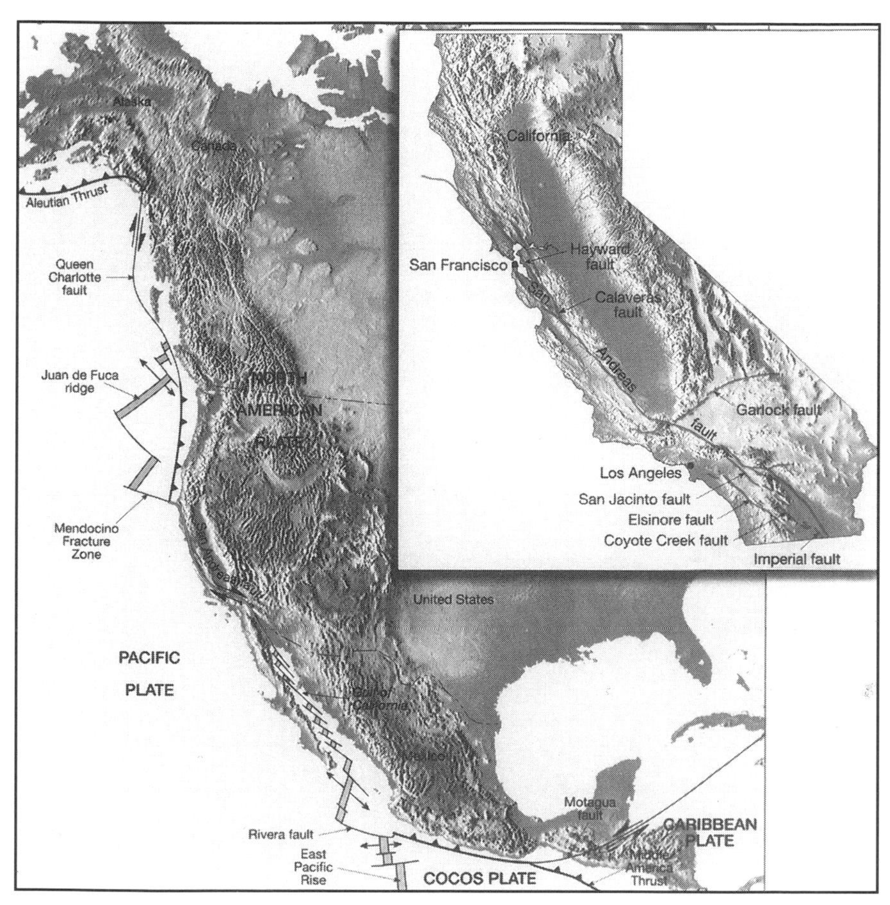

Fig. 50. San Andreas Fault. Photograph from Frederick K. Lutgens and Edward J. Tarbuck, *Essentials of Geology*, 6th ed. (Upper Saddle River, N.J.: Pearson Education, Inc., 1998), 355; reprinted with permission.

CHAPTER FIVE

French Fantasies

In the latter half of the seventeenth century and the first half of the eighteenth century, French cartographers rose to a level of prominence. Nicolas Sanson, Guillaume de l'Isle, Nicolas de Fer, Jean Nicolas Bellin, and Jean Baptiste Bourguignon d'Anville all produced significant maps. They not only published the most inclusive and up-to-date North American maps of the times but introduced new geographic facts as they were uncovered in that region. One of the specific areas of interest to the French cartographers was the Great Lakes, on which the French Canadians played major roles as discoverers. Among the many positive contributions that were generated by these French cartographers, however, one apparently minor inappropriate introduction of a geographic entity has been memorialized in the archives of the United States of America.

The smallest of the French fantasies that were perpetrated on North America was the insertion of two nonexistent islands on the largest of the Great Lakes. Although the genesis of the fantasy has not been resolved, a credible solution has been offered. The apocryphal islands were eventually deleted from maps, but the name of one of the islands remains a specific reference point in the definition of the boundaries of the United States of America.

The Great Lakes are arguably the most important geographic entity in the United States. They cover over 95,000 square miles and contain over 67 trillion gallons of water, making them the largest body of fresh water in the world. Their connecting waterways constitute the world's largest inland water transportation system. The distance from the westernmost port of Duluth on Lake Superior to the exit from Lake Ontario into the St. Lawrence River is 1,160 miles, and an additional passage of 1,182 miles down the river brings a ship into the open seas of the Atlantic Ocean.

About 500 million years ago, the start of modern geological time, the current area of the Great Lakes was covered with molten lava that rose from the earth's core. After the lava subsided, a shallow sea developed. About one million years ago, glaciers began to spread over the region and sculpt the lakes. Five periods of glacial spread and retreat have been defined. The glacial effect was weight crushing down and compacting the loose crust of the earth. The glaciers slowly moved southward, gouging the land to varying depths.

With each southward movement of the glaciers, the sun warmed the earth and the rate of melting exceeded the rate of movement, and the land debris built up and formed the shore line of a lake. In the first stage of formation of the Great Lakes' region, two small lakes formed in the regions of Chicago and Toledo, and drained via the Chicago and Maumee Rivers into the Mississippi River. As the glacier melted, Saginaw Bay formed in western Lake Huron, and an outlet developed in the Grand River Valley near current Flint, Michigan. In time, Green Bay was formed, draining by a river system into the Mississippi and, as the ice continued to melt the water in the area of current Lake St. Clair, the water mass spread south to form Lake Erie.

At the next stage of glacial movement and warming, the Finger Lakes, draining down the Susquehanna River into Chesapeake Bay, were formed in western New York before Lake Ontario was defined. Lake Erie increased in size, and the southern portion of Lake Huron was uncovered. Later, the upper area of Lake Michigan melted and the western region of Lake Superior became a body of water that drained through the St. Croix River into the Mississippi River. During the next stage, the three upper lakes, Superior, Michigan, and Huron, joined to become a large body of water draining via the Ottawa River into the St. Lawrence River. Initially, when the St. Lawrence Valley was cleared to a depth below sea level, salt water flowed all the way from the ocean as far west as Toronto and Hamilton, Ontario. At that point in time, the waters of Lake Ontario emptied by a narrow River down the Mohawk Valley. About 25,000 years ago, Niagara Falls was born from an escarpment that was buried below the spreading waters, the St. Lawrence River was formed to drain Lake Ontario northeastward to the ocean, and Lake Ontario contracted its shoreline, no longer draining into the Mohawk Valley.

All of the lakes except Erie have depths that are well below sea level. Lake Superior's maximum depth of 1,333 feet is 732 feet below sea level; Lake Michigan's maximum depth of 925 feet is 346 feet below sea level; Lake Huron's maximum depth of 750 feet is 173 feet below sea level; Lake Erie's maximum depth is only 210 feet, and Erie is the only one of the Great Lakes with its maximum depth *above* sea level (358 feet); Lake Ontario's maximum depth of 804 feet is 559 feet below sea level.

The exit of the salt water from the St. Lawrence Valley and the creation of the St. Lawrence River made possible the exploration that began in 1535. The St. Lawrence River became the gateway to the exploration of the Great Lakes. Between 1534 and 1541, Jacques Cartier sailed on three occasions for France in hopes of finding a Northwest Passage, which had escaped Verrazzano during his 1524 voyage. In 1534, Cartier left Saint-Malo, the leading port in Brittany, and made landfall at Cape Bonavista, Newfoundland. He sailed in the Gulf of St. Lawrence, discovering the Magdelan island group. He entered the Baie de Chaleur, anchored in Gaspé Bay, and sighted Anticosti Island before returning home.

Almost immediately after Cartier's return, King François I of France ordered a second voyage. Cartier departed Saint-Malo in May 1535 as commander of a fleet of three vessels, *La Grande Hermine*, *La Petite Hermine*, and a small pinnace, *L'Emerillon.* When Cartier anchored on August 10, he named the large bay for Saint Lawrence, who was honored on that date. The fleet entered what the Native companions, who had joined him during his first voyage, called the Great River of Hochelaga (Montreal). In Cartier's *Bref Récit* of his journey, the river is referred to as "le chemin du Canada," (the road of Canada) the first appearance of the word "Canada" in European literature.

Cartier reached Tadoussac, a Native settlement, at the mouth of a river, named Saguenay by the Natives, that entered the St. Lawrence River from the north. Continuing upriver, Cartier discovered and named a large island, Orléans, in honor of the king's son, the Duc d'Orléans. In search of a passage to the Indies, Cartier used the smaller pinnace to reach the area of current Montreal, which, at the time, was the site of a fortified Native village known as Hochelaga. The travelers were unable to continue westward because of a series of rapids, which La Salle would later name "La Chine" rapids because they had been thought to be a gateway to the Orient. Cartier and his men wintered at the mouth of the St. Charles River near Quebec and sailed back to France in May 1536.

Cartier's third expedition left Saint-Malo May 23, 1541, with the charge of establishing a colony on the banks of the St. Lawrence River in order to have a base for future incursions into Canada, which Cartier referred to as "La Nouvelle France." They built a settlement on Cap Rouge, naming it Charlesbourg-Royal. After going upriver to the site of current-day Montreal and returning to his settlement, Cartier wintered at Cap Rouge and returned to France in the spring of 1542.

The imagery of the Great Lakes, and particularly the greatest of the Great Lakes, Superior, began almost immediately, but was initially a product of imagination based on the tales of the Native Americans. After the voyages of Cartier, suggestions of the presence of a great inland lake or lakes in the North American continent were depicted on sixteenth-century maps. De Jode's map "Americae Pars Borealis, Florida, Baccalaos, Canada, Corterealis" (see figure 28), published in 1593, includes a "Lago de Conibas" in the middle of the continent, connected to a Northwest Passage. The first atlas devoted to the Americas, *Descriptionis Ptolemaicae Augmentum*, published by Cornelis Wytfliet in 1597, included a map centered on "Lago de Conibas" (see figure 29). On Edward Wright's map from about 1599, a suggestion of a Great Lake appears as "The Lake of Tadousac" (see figure 31).

Cartier's three voyages set the stage for Samuel de Champlain, the first person to discover and map any part of the Great Lakes system. Champlain, anointed "Father of New France" by the distinguished historian Samuel Elliot Morison, was born in the coastal town of Brouage in the district of Saintonge (Charente-Inférieure) around 1570. In 1603, shortly after returning from the West Indies, Champlain participated in a trading expedition led by François Gravé, Sieur du Pont. The small fleet sailed up the St. Lawrence River to Tadoussac, at the mouth of the Saguenay River, where active fur trading with the Native Americans had been established. Champlain chronicled his travels up the St. Lawrence River in *Des Sauvages, ou, Voyagès de Samuel Champlain, de Brouage, fait en la France nouuelle, l'an mil six cens trois* . . . , published in Paris in 1603. In the text, the name "Kébec" appeared for the first time, applied to the bend of the river, where Quebec is now located. "Quebec," which derives from a Micmac word meaning narrows, first appeared in its present form on a 1601 map by Levasseur. Champlain sailed upriver to the site of Montreal and on to the Lachine Rapids, which he named Sault Saint Louis. During his travels in 1603, he received a description of the Ottawa River and Lake Ontario from the local Native Americans, who also referred to Niagara Falls and other large lakes. Champlain arrived back in France on September 20, 1603.

On April 7, 1604, Champlain left France on an expedition led by Pierre du Gua, Sieur de Monts. In May, Champlain commanded a small pinnace as it explored the Acadian coast and the Bay of Fundy, which he named La Baie Françoise. He also named and charted the St. John River and later entered Passamaquoddy Bay and continued up the Sainte-Croix River, where he established a settlement on a midstream island. In September 1604, in command of a small vessel with twelve sailors and two Natives, Champlain left the island to explore

"Norumbega," the early name for New England. He named the Manan Islands and Mount Desert Island, and, after exploring the harbors of Mount Desert Island, he sailed up Penobscot Bay.

After wintering at Sainte-Croix, Champlain took off, in June 1605, with du Monts, and sailed up the Kennebec River, reaching the current site of Bath, Maine. Champlain charted Saco Bay and, in July, passed Cape Ann and entered Boston Bay. He named the Charles River "Rivière du Gua," after the family name of the Sieur de Monts. Shortly thereafter, the vessel anchored in the same harbor that the Pilgrims would enter fifteen years later. While sailing along the coast of Cape Cod, Champlain named the land Cape Blanc. Some of the men went ashore and met with the Natives on Nauset Beach. On Champlain's return to Sainte Croix, the buildings were disassembled, and the base camp was moved to Port Royal in the Annapolis Basin of Nova Scotia. From March through November 1606, Champlain made his third and last voyage along the New England coast.

In early September 1607, Champlain left Acadia for France, never to return to that area. A manuscript map, "Les Côtes et Grandes Isles de la Nouvelle France," dated 1607, resides in the Library of Congress. The map is the first delineation of the coast from Cape Sable to south of Cape Cod. Marc Lescarbot's 1609 printed map antedates Champlain's first printed map by three years, and depicts the St. Lawrence River to Hochelaga (Montreal) and the unnamed Lachine Rapids. There is no evidence of any Great Lake. The first map to show any segment of the Great Lakes was a manuscript drawn in 1611 by Velasco for James I of England. On that map a recognizable representation of Lake Ontario appears, doubtlessly obtained from Natives' accounts.

Champlain briefly returned to France on September 28, 1607, and on July 3, 1608 he was back in Quebec where he formally claimed the land for the King, and thereby gave a certificate of birth to the city and France's empire in North America. Champlain and his twenty-three accompanying men wintered at Quebec; only Champlain and seven others survived. In June 1609, Champlain began his first incursion against the Iroquois. At the end of July, he led a group to the lake on which he bestowed his name. At that locale, the French defeated a larger contingent of Mohawks, and Champlain memorialized the event with a drawing that included himself as the central figure. The sketch was included in his *Les Voyages* of 1613.

Champlain returned to France in 1610, married, and was back in Quebec in May 1611. In the fall of 1611, he once again returned to France, where he remained for almost twenty months. In the spring of 1613, Champlain undertook his first western exploration in

Canada. The goal was to find a water route to the west and also to open the interior to fur traders. After portaging two canoes around the Lachine rapids, the party, consisting of four other Frenchmen and a Native guide, traveled along the Ottawa River. The description of the travels is covered in Champlain's *Quatriesme Voyage*, which was planned as a separate book but was published as part of *Les Voyages* as a single volume in 1613.

The first printed map to include any part of the Great Lakes is the one that was published in 1613 as part of *Les Voyages du Sieur de Champlain*. The map, titled "Carte Geographiqve de la Novvelle Franse Faictte par Le Sieur de Champlain Saint Tongois Cappitaine Ordinaire pour Le Roy en La Marine facit len 1612," includes an unnamed lake containing the legend "Lac Contenant 15 Journees de Canaux des Sauuges" where Lake Ontario is located. A narrow waterway leads westward from that lake into part of a large body of water bearing the text "Grand Lac Contenant 300 Lieux de Long" (see figure 51).

A second map, titled "Carte geographique de la Nouelle franse en sonnvraymeridiein," is also included in *Les Voyages* of 1613. It covers only the easternmost part of Lake Ontario, which bears the name "Lac St. Louis." The information on which the two maps were based was derived from three accounts, accompanied by sketch maps, which were gathered by Champlain from Natives during his stay in New France, in 1603. One of the descriptions offered by the Natives extended from the Lachine Rapids to Lake Huron, while the other two went as far as Lake Erie. This first cartographic evidence of a portion of the Great Lakes was based solely on hearsay.

The first viewing of a Great Lake by a European occurred in the summer of 1615, when Champlain, accompanied by two Frenchmen and ten Natives, paddled up the Ottawa and Mattawa Rivers to Lake Nipissing, and then followed the French River into Georgian Bay on Lake Huron. Champlain assigned the name "La Mer Douce" to the lake as he traveled along the eastern shore of Georgian Bay. While in the land of the Hurons, Champlain dispatched one of his interpreters, Etienne Brûlé, to the western end of Lake Ontario and the Susquehanna River. Brûlé crossed the Niagara River and, during the winter of 1615–16, explored the Delaware River to its mouth on the Atlantic coast. Brûlé returned to Huronia, and the following winter he traveled along the northern shore of Lake Huron. Meanwhile, Champlain led a group of five hundred Native warriors to the Bay of Quinte at the eastern end of Lake Ontario. They crossed the lake to Stoney Point in New York via a string of islands near the eastern shore of Lake Ontario. The group attacked an Iroquois village; during the battle, Champlain was injured.

A narrative of Champlain's explorations and activities between 1614 and 1618 is presented in his *Voyages et descouvertures faites en la Nouvelle France depuis l'année 1615 iusques à la fin de l'année 1618*, which was published in Paris in 1619. No map appears in the work, but Champlain probably intended to include one. The only known example of an incomplete printed map, which bears the inscription "facit par le Sr de Champlain: 1616" resides in the John Carter Brown Library in Providence, Rhode Island. This is the first map to depict part of the Great Lakes based on European exploration. "Lac St. Louis" (Lake Ontario) is presented quite accurately. A waterway (Niagara River) connects "Lac St. Louis" with an unnamed small narrow body of water that represents Lake Erie. The western portion is widened in the region of Lake St. Clair, which, in turn, connects with a large "Mer douce" (Lake Huron). A broad strait (Mackinac) connects "Mer douce" with an unnamed lake (Michigan). The same plate that produced the 1616 map was used by Pierre Duval in the 1653 production of his map (see figure 52).

Champlain's explorations in North America ceased after 1616. Between 1616 and the time of his death in 1635, he shuttled back and forth between Canada and France several times. In 1620, Chaplain was accompanied to Quebec by Hélène, his wife of ten years, whom he had married when she was twelve years old. She remained in Quebec for four years before happily returning to Paris. After a short stay in France, Champlain was back in Quebec in 1626. In 1628, England declared war on France and Quebec, and blockaded the St. Lawrence River, cutting off all supplies. In July 1629, a landing force, led by David Kirke, took possession of the French settlement and raised the English flag. An armistice was declared that fall, and the Treaty of St. Germain-en-Laye, signed on March 29, 1632, returned Canada and Acadia to France.

In 1632, Champlain's *Les Voyages de la Novvelle France occidentale, dicte Canada, faits par le Sr de Champlain*, was published in Paris. He made his last voyage to Canada in 1633, and died at Quebec on December 25, 1635. He was buried in the city he founded on the grounds of Notre Dame de la Recouvrance.

Champlain's last map, "Carte de la nouuelle france . . . 1632" was drawn specifically for *Les Voyages* of 1632. The map provides a graphic summation of the knowledge of the Great Lakes up to the time of the temporary expulsion by the English. Significant changes occurred in the representation of the Great Lakes. "Lac St. Louis" (Lake Ontario) is more appropriately shaped, while the waterway extending westward from Lake Ontario into "Mer douce" (Lake Huron), including an unnamed and poorly portrayed Lake Erie. "Mer douce" contains a legend that states: "Descouvertures de ce grand lac, et de toutes se terres depuis le Sault St Louis par le Sr de Champlain es années 1614 et 1615 iusques en l'an 1618." The large lake has a reduced bay where Saginaw Bay is now located, and that bay contains a large fictitious island. "Grand lac" is shown to the west of "Mer douce," and a "Sault de Gaston," which is thought to be Sault Ste. Marie, connects the two large bodies of water. Most scholars have concluded that "Grand lac" represents Lake Superior. C. E. Heidenreich, Professor at York University, Ontario, Canada, however, suggests in his article "Explorations and Mapping of Samuel de Champlain, 1603–1632" that Champlain attempted to conflate several verbal reports that he received after 1616, and that he confused Lakes Michigan and Superior. The small body of water emptying into what is possibly Lake Michigan rather than Superior has the word "Puan," referring to Green Bay on the shore. Champlain erroneously placed the "Sault" between Lakes Michigan and Huron (see figure 53).

In 1615, four Recollect Fathers, Joseph La Caron, Denis Jamet, Jean d'Olbeau, and Lay Brother Pacificus Du Plessis, accompanied Champlain to New France, thereby initiating the role that priests would play in the history of the Great Lakes. In 1623, Father Gabriel Sagard Théodat arrived in Quebec and proceeded to Huronia, the name given to land controlled by the Hurons, and the shores on Lake Huron, which he described in *Le Grand Voyage du pays des Hurons*, published in 1632. In 1625, the Recollects invited Jesuits to join them in their efforts to convert the Native Americans, and, that year, Jean de Brébeuf arrived in Quebec. In 1634, he accompanied Jean Nicollet, who was dispatched by Champlain to define a passage that would eventually provide access to the Orient. Nicollet's wardrobe on his journey included clothes suitable for calling on the Emperor of Cathay. Nicollet established a base in Huron country, and set forth on his trip. He and several companions canoed on Lake Huron, crossed Potaganissing Bay at Drummond Island, and continued through the Straits of Mackinac. The group entered Lake Michigan, and Nicollet became the first European to view that lake. They proceeded to Green Bay, and entered the Fox River, where they met with the Winnebagos. They paddled the length of Lake Winnebago and the upper reaches of the Fox River, and ended at a narrow strip of land between the Fox and Wisconsin Rivers before returning to their home base. Nicollet was convinced that only three more days of westward travel from his terminus would have brought him to the great sea that offered access to the Orient.

The period of Champlain's contributions to the exploration and mapping of the Great Lakes that depicted the concept of a series of large lakes to the west was followed by four decades dominated by the activities of the Jesuits, referred to as "the Black Robes." Initially impeded by the trade protectionism of the Hurons, their travels and establishment of missions eventually extended over the shores of many of the lakes. The courageous Fathers chronicled their activities in the *Jesuit Relations*, which was published annually from 1610 through 1791. Each year, the Fathers at the missions submitted reports that were edited by the Superior at Quebec and published in Paris by the firm of Sebastian Cramoisy. The activities of the *coureurs de bois*, who, in their quest for furs, were the true pioneers of discovery, were incorporated in these works. By contrast to the vague speculation that Etienne Brûlé *may* have reached the shores of Lake Superior in 1622, the *Relations* provided documentation of events as they occurred..

The initial period of Jesuit activity began with their arrival in 1625 and came to an abrupt end in 1649 when the Hurons, which was the group with whom the priests lived and to whom they preached, were defeated by the Iroquois and forced to disperse westward to the Wisconsin region, abutting the Sioux. Most of the early activity of the Jesuits centered in Huronia, the land of the Hurons on the eastern shores of Lake Huron at Georgian Bay. During the early 1640s, Lake Huron, Lake Superior, and the northern shore of Lake Erie were surveyed more extensively. Father Vimont's *Relation* of 1641 contains the first mention of a river that serves as an outlet for Lake Huron. This river, according to the report, flows into Lake Erie, known as the Lake of the Cat Nation, and then enters the territory of the Neutral Nation, where it leaves Lake Erie as the Ongiaaha River (Niagara) that empties into Lac. St. Louis (Ontario). In 1640–41, Fathers Jean de Brébeuf and Gabriel Chaumonont viewed Niagara Falls. Fathers Jogues and Rayambault built missions at St. Ignace on the Straits of Mackinac and on the shore of the Fox River, and were the first to reach and describe Sault Ste. Marie. These accomplishments were reported in Father Jérôme Lalemant's *Relation* of 1642, in which it is stated: "They started from our house at Ste. Maries, the water that bathes the land of the Hurons, they reached the Sault . . . ." At Sault Ste. Marie, the Jesuits met the Ojibways, who informed them that an eighteen-day journey west across Lake Superior would bring them to an Indian Nation known as the Nadouissuex or Sioux, and that a great river flowed in that region. "The first nine days are occupied in crossing another great Lake that commences above the Sault . . . ." Fathers Brébeuf and Lalemant built a mission and fort at Matchedash Bay at the south end of Georgian Bay and named it "Ste. Marie." When the area was captured by the Iroquois in 1649, the priests were killed, becoming the first martyrs in the region.

In Father Ragueneau's *Relation* of 1647–48, all of Lake Huron is described and Lake Superior is mentioned for the first time by name, "Ce Lac Superieur." The position of Lake Superior relative to Lake Huron and Lake Michigan, called "Lac de Puants," is explained. That name was applied to the lake and to the bay that eventually became Green Bay because the Puant Indians in the region said they came from the shores of a distant salt water sea and were referred to as the tribe of the stinking water (puan). By the end of the 1640s, the Jesuits, based on their own travels and the descriptions of Native informants, had constructed a rough picture of the size and relationships of the Great Lakes.

Nicolas Sanson's 1650 map (see figure 54) was dependant on Jesuit reports, and was the first printed map to depict all five Great Lakes in a recognizable form. Erie appears as "Erie ou Du Chat." The name was first applied by Boisseau in 1643 as "Lac Derie" to a part of a stretch of water shown on the Champlain map. In 1656, Sanson published another map that presented the Great Lakes in more detail. The eastern lakes are shown in reasonably true form. In 1657, a printed map attributed to Francisco Bressani appeared. An inset portrays the martyrdom of Fathers Brébeuf and Lalemant (see figure 55). The two extant copies of the map, entitled "Novae Franciae Accurata Delineato 1657," are in the Bibliothèque Nationale de France in Paris and the National Archives of Canada in Ottawa. On the map, Lake Ontario is named "Lactus Sanct Ludovici"; Lake Erie is called "Enyat"; and "Mar Duce" encompasses Lakes Huron, Michigan, and Superior.

When the mission of Ste. Marie at the south end of Georgian Bay was destroyed by the Iroquois in 1649, the Jesuits moved west. They quickly built four new missions, one at Sault Ste. Marie; one at La Pointe on Chequamegon Bay, three hundred miles east of the Sault on the south shore of Lake Superior; one at St. Ignace on the Straits of Mackinac; and the mission of St. Xavier at the mouth of the Fox River on the west shore of Lake Michigan. In 1654, peace between the Iroquois and the French opened the eastern Great Lakes to more exploration. Father Le Moyne's mission to the Onondaga that year was the first of several journeys to the land of the Iroquois and around Lake Ontario. The peace also allowed the missionaries and *coureurs de bois* to extend the travels in the region of the western lakes.

Two of these *coureurs*, Médart Chouart, Sieur des Grosseillers, and his brother-in-law, Pierre-Esprit Radisson, traveled extensively in the Great Lakes region beginning in 1654. They are purported to have set forth from Three Rivers in 1658, to have gone up the Ottawa River, across the Nippising, and into Georgian Bay. They allegedly followed the shoreline around Lake Huron, camped on the Manitoulin Islands, and crossed Lake Michigan to Green Bay, where they went up the Fox River. They crossed over the watershed at Portage and may have entered the Mississippi River fifteen years before Marquette and Jolliet. After traveling over Lake Michigan, Grosseillers and Radisson were supposed to have reached Sault Ste. Marie, paddled up the St. Mary's River, entered Lake Superior, and become the first Caucasians to traverse that lake. They wintered at Chequamegon Bay in 1659–60, crossed the land in the corner of current Wisconsin, and came upon the Mississippi River again.

The two *coureurs de bois* left no reports, and it is impossible to filter fact from fiction. The discovery of the Mississippi by Grosseillers and Radisson is disputed, and it is contended that the river they noted was not the Mississippi. The earliest reference to a great river south of the Great Lakes appeared in the *Relation* of 1661–62 by Father Jérôme Lalemant. The first mention of what is unequivocally the Mississippi is in the *Relation* of 1666–67, where Father Claude Allouez, writing about the Sioux, states: "These are people dwelling to the west of this place (Chequamegon Bay) toward the great river, named Messipi." In 1665, Father Claude Allouez was sent to Lake Superior where he established a mission at Pointe du Esprit, and included the first reliable description of the south shore of Lake Superior in the *Relations* of 1666–67. That issue of the *Relations* also contains the first mention of Lake Michigan. By 1670, all of the Great Lakes except the east shore of Lake Superior and Lake Michigan and the west shore of Lake Huron were thoroughly explored and charted.

In the 1669–70 *Relation,* Allouez includes a detailed description of Lake Superior and the second reference to the Mississippi River by name. In referring to the Natives along the Fox River, he writes: "Their River leads by a six days' Voyage to a great River named Messi-Sipi." In 1671, Dablon was made Superior General of the Jesuit missions in New France, and was responsible for the *Relations* of 1670–71, published in 1672. The edition includes the first printed map dedicated to Lake Superior, which would become the locale of the islands of fantasy, the focus of this chapter.

Father Dablon was the New France Jesuit most interested in geography, but other priests were versed in sciences applicable to mapmaking. Bressani was a mathematician; Gallinée was an astronomer. The general crudeness of the maps was related to the tools available. All that the priest-cartographers had to work with were the compass, cross staff, back staff, and astrolabe. They were aware that the compass deviated from true north, and tables were used for correction. Measurement of latitude required declination tables, and parallax had to be corrected. Measurement of longitude was more difficult and less precise. It could be calculated if both the angle of travel and the latitude were known, by applying the Pythagorean theorem, or by a table of tangents if the latitude and compass direction was known, or using navigation tables. At the time, longitude was also measured by timing the eclipse and determining the precise time difference between the North American location and a specific point in France.

The 1672 map entitled "Lac Svperievr et avtres lievx ou sont les Missions des Peres de la Compagnie de Jesvs . . ." (see figure 56) is one of the most impressive maps of New France published in the seventeenth century. Its delineation of all of Lake Superior and the northern portions of Lake Michigan and Lake Huron was unmatched until formal surveys were made in the nineteenth century. It was the first map to distinguish between the "Baye des Puans" (Green Bay) and "Lac des Ilinois" (Lake Michigan). Isle Royale in Lake Superior is named "I. Minong," and there is no other name assigned to an island within Lake Superior.

Another order of priests, the Sulpicians, who first came to New France in 1635, also conducted a series of explorations which contributed to the knowledge of the Great Lakes. One of their expeditions was led by François Dollier de Casson. accompanied by Father Bréhant de Galinée and René-Robert Cavelier de La Salle. In 1669–70, the trio became the first Europeans to enter the Niagara River from Lake Ontario, the first to winter on the shores of Lake Erie, and the first to make reasonable surveys of the shore of both lakes. Galinée, who was trained in mapmaking, drew a map; the original is lost, but a reduction appears in volume 3 of Nicolas Faillon's *Histoire de la colonie française en Canada*, 1685. Based on personal observations, Galinée drew the south shore of Lake Ontario, the north shore of "Lac Erie," and the east and north shores of "Lac des Hurons," while some of the other shorelines were drawn as unknown. Most of the lakes are shown in very distorted forms.

In 1670, Jean Talon returned to New France for his second term as Intendant and immediately initiated efforts to expand exploration with four major goals: to increase the fur trade, to search for ore that could be mined, to discover a waterway to the Orient, and to stake claims of possession in the name of the King of France.

On June 4, 1671, Simon F. Daumont, Sieur de Saint-Lusson, conducted a vain search for ore in the Lake Superior Region and also failed to discover a waterway to the west. The goal of economic expansion followed a parallel course with the proselytizing by the Black Robes. In the presence of Louis Jolliet and four Jesuits, Claude

Dablon, Superior of the Missions of the Lakes, Gabriel Druilletes, Claude Allouez, Louis André, and many Native tribes, Saint-Lusson formally proclaimed, as reported by Francis Parkman in his *La Salle and the Discovery of the Great West*:

> In the name of the Most High, Mighty, and Redoubted Monarch, Louis Fourteenth of that name, Most Christian King of France and of Navarre, I take possession of this place, Sainte Marie du Saut, as also of Lakes Huron and Superior, the Island of Mantoulin, and all countries, rivers, lakes, and streams contiguous and adjacent there unto; both those which have been discovered and those which may be discovered hereafter, in all their length and breadth, bounded on the one side by the seas of the North and of the West, and on the other by the South Sea . . . .

Père Jacques Marquette continued Father Allouez's work at Lake Superior until the mission was threatened by the Sioux. Then, in 1671, Marquette left to establish a mission at St. Ignace on the north shore of the Straits of Mackinac. In May 1673, he left with Louis Jolliet, who was traveling with Intendant Talon's encouragement, with the hope of finding the much talked-of Mississippi River, and of determining where it discharged its waters. Accompanied by five *voyageurs*, Marquette and Jolliet crossed Lake Michigan to Green Bay and the Fox River, following the route of Jean Nicollet. After ascending that river, the group portaged to the Wisconsin River, and followed the Wisconsin River to its entrance into the Mississippi. They descended the Mississippi River to the point where the Arkansas River enters from the west, and then returned to their starting point.

As the last quarter of the seventeenth century began, all that remained to complete the knowledge of the contour of the Great Lakes was exploration of Lake Erie, which had been considered off-limits because of the threat of the warring Iroquois. Once the war between the Iroquois and the Huron ended, the lower Great Lakes—Ontario and Erie—became safe to travel. Niagara Falls on the Niagara River was known to exist, but had not been described by a European based on direct viewing. These achievements were part of the accomplishments of La Salle and his three lieutenants: Henri de Tonty, La Motte de Lussière, and Father Louis Hennepin.

René-Robert Cavelier de La Salle, who forsook his position as Brother Ignatius in the Society of Jesus, arrived in New France in 1667, at age 23, determined to explore the region around the Mississippi River and exploit it for trade. On his first journey, in 1669, he started at the Lachine Rapids with twenty-three men and embarked on Lake Ontario. After crossing the lake and visiting a Seneca village along the banks of the Genesee River, La Salle coasted the south shore of the lake to the mouth of the Niagara River. At the head of Lake Ontario, he met up with Louis Jolliet. The Sulpician priests who accompanied La Salle proceeded to Lake Erie and, after wintering there, passed through the strait by which Lake Erie joins Lake Huron, landing near current Detroit. This is the first recorded passage of white men through the Strait of Detroit, although Jolliet probably passed through that strait on his return from the Upper Great Lakes. The Sulpicians continued along the east shore of Lake Huron and on to Sault Ste. Marie, where they met Fathers Dablon and Marquette before returning to Montreal.

In 1677, La Salle was authorized by Jean-Baptiste Colbert, Governor of New France, to build boats and navigate the Great Lakes and the Mississippi River. La Salle's companions on his historic journey westward were most diverse. La Motte de Lussière came from a noble French family and helped finance the venture. Henri de Tonty was an Italian who had lost his hand in a battle in Sicily; he wore an iron substitute in his glove, which accounted for his nickname, "silver claw." Father Louis Hennepin was a Recollect priest, whose books, replete with fantasies, were to become the most widely read works of French exploration in North America.

In 1678, Father Hennepin went up the Niagara River and became the first European to see and describe the Niagara Falls. The first printed depiction of the natural wonder appears in his *Nouvelle découverte d'un très grand pays,* which was published in 1697. After building their vessel, the *Griffon,* La Salle and his companions left the head of the Niagara River and sailed across Lake Erie, where no ship had previously sailed. The group continued through the Detroit River and Lake St. Clair, named for the blessed abbess of Assisi, because they entered it on her day, August 12. The vessel entered Lake Huron, sailed into the Straits of Mackinac, and anchored in the bay of St. Ignace, near the chapel built eight years previously by Père Marquette, whose bones, having been found two years earlier on the eastern shore of Lake Michigan, were buried under the floor. La Salle then sailed across Lake Michigan to Green Bay, where the ship, loaded with furs, was sent back east, but it was lost at some unknown location on the return voyage.

In 1678–79, Greysolon Dulhut (Duluth) explored the western portion of Lake Superior, while Tonty paddled down the eastern shore of Lake Michigan. The knowledge of the Great Lakes was essentially complete when La Salle and his companions reached the mouth of the Mississippi River on April 7, 1682, and took possession of the surrounding land for France, naming it "Louisiane" for the King. Five years later, La Salle sailed from France, but failed in his attempt to find the mouth of the Mississippi River on the Gulf Coast. La Salle was murdered near the Trinity River, at age forty-three, by his mutiniering men. His wrecked vessel has recently been discovered along the Gulf Coast and parts have been raised.

As he prepared for his explorations, La Salle probably used Nicolas Sanson's 1656 expansion of the 1650 map showing the five Great Lakes (see figure 54). The 1670 map by Galinée concentrating on the eastern Great Lakes and the 1672 map of Lake Superior also afforded valuable information. During the remaining three decades in the seventeenth century, most of the maps focusing on the Great Lakes were related to the travels of Jolliet and La Salle. Between 1670 and 1684, manuscripts by Louis Jolliet, Jean-Baptiste-Louis Franquelin, and Abbé Claude Bernou dominated the production, but little in the way of new information was depicted. Louis Jolliet's "Nouuelle Decouuerte de plusieurs Nations dans la Nouuelle France en l'année 1673 et 1674," generally provides a distorted depiction of the Great Lakes. Louis Hennepin's "Carte de la Nouuelle France et de la Louisiane," published in 1683, presented the word "Louisiane" for the first time on a printed map. On that map, Lake Ontario is named "L. Frontenac." "Lac de Conty ou Erie," Lac d'Orleans ou Huron," "Lac Dauphin ou Illinois," and "Lac de Conde ou Superieur" are the names assigned to the other four Great Lakes.

In 1684, Franquelin became the official cartographer for La Salle and produced his first of two original maps; only copies have survived. They depict the Great Lakes more realistically than earlier maps. In 1686, Franquelin was appointed Hydrographe du Roy at Quebec, and continued producing manuscript maps of the Great Lakes region. In 1688, Vincenzo Coronelli, a Venetian Franciscan who served as official cartographer to Louis XIV, published "Partie Occidentale de Canada ou de la Nouvelle France," which shows the Great Lakes and La Salle's explorations. It was the first major map of New France to be published subsequent to Sanson's 1656 map. Hennepin's 1697 "Carte d'un tres grand Pays entre la Nouveau Mexique . . ." presents a concept of the Great Lakes at the end of the seventeenth century.

The eighteenth century opened with the settlement of the first American city on the Great Lakes waterway when Antoine de la Mothe-Cadillac received permission to build a frontier post at Detroit. Once the fort was completed, Cadillac sent for his wife and other women, who, in the spring of 1702, sailed up the St. Lawrence River, then across Lake Ontario, portaged to Lake Erie, and finally arrived at Detroit; they became the first white women to reach a western French post.

Early in the eighteenth century, far from the Great Lakes, the geography of which had been established, the first of the French fantasies related to North America appeared on a map. A so called "Rivière Longue" (River Long), was depicted as a significant, albeit apocryphal, waterway in a book entitled *Nouveaux Voyages de Mr le baron de Lahontan dans L'Amérique Septentrionale,* which was published simultaneously in French and English editions in 1703. The book, which chronicled the travels of Baron de Lahontan in North America, was subsequently translated into German and Dutch, and became the most widely read travelogue pertaining to North America at the time.

The author, Louis-Armand de Lom d'Arce, was born in France, in the village of Lahontan in the frontier department of Basse-Pyrenées, on June 9, 1666. When he was eight, his father died at the age of eighty, and Louis inherited the title of Baron de Lahontan et Hesleche. As a youth, the Baron was enrolled as a cadet in the famous Bourbon regiment, and later transferred to the marine corps responsible for the care of the colonies in North America.

In 1683, as part of a contingent of marines, Lahontan left for Canada, and was billeted at Beupré, about seventeen miles downriver from Quebec. He hunted in the region with the Natives and learned their language. In 1684, he proceeded to Montreal and participated in a mission against the Iroquois, crossing Lake Ontario to the Famine River, where a peace conference was held. In September 1687, he took part in Governor Denonville's campaign against the Seneca Indians in the region of current Rochester, New York. Later, although Lahontan petitioned to return to France in order to untangle his desperate financial affairs, he was sent to the distant post of Fort St. Joseph, on the strait that joined Lakes Erie and Huron, where he met Dulhut, for whom Duluth would be named. Lahontan assumed the position of commandant of the small post.

In the spring of 1688, Lahontan set out with most of his men for the French military and trading station on the north shore of the strait of Mackinac. After traveling as far west as Sault Ste. Marie, he returned to Fort St. Joseph. Because he assessed the fort as being of little value, he destroyed the blockhouse and stockade, and moved his entire contingent to Mackinac.

At this point in the narrative of *New Voyages*, Lahontan relates his apocryphal journey to the "Rivière Longue." The description of the journey appears in the book in Letter XVI, dated at "Missilimackinac,"

Royale," "I. Maurepas," "I. Beauharnois," "I. Hocquart," and "I. Ste. Anne" (see figure 60).

A series of events occurred subsequent to the 1725 map that explain the inclusion of the two fanciful islands and the names assigned to the real islands on the 1735 map by the elder Léry. Reports of copper deposits in the Great Lakes region go back to tales of the *coueurs de bois* to Samuel de Champlain. Mining of valuable ore was continually regarded over the years as having great potential for the economy of New France. In 1729, the newly appointed Intendant of New France, Gilles Hocquart, who would hold the position until 1748, moved to improve the economic status and commercial enterprises of the French in North America. Intendant Hocquart and Charles Beauharnois de la Boishe, Governor General of New France from 1726 to 1747, issued an order in 1729 to investigate the mining potential on Lake Superior.

In 1731, Louis Denys de La Ronde, who was appointed by Governor Beauharnois, took command of a fur post on Madeline Island at Chagouamigon on the southwest shore of Lake Superior. "Isle La Ronde" is the name assigned to current Madeline Island on the 1735 Léry chart. La Ronde built two ships to ply the waters of the lake in order to discover copper deposits and eventually mine the region. A Montreal merchant, Louis Charley Saint-Ange invested over 50,000 livres in the venture. In 1735, La Ronde submitted a report to Hocquart and Beauharnois indicating that he had not uncovered the copper deposits that the Natives had reported. The disappointment was reaffirmed in reports made during the three subsequent years. In 1739, German mining consultant sent by the Comte de Maurepas, put an end to the enterprise when they concluded that mining the region was impractical because of the potential of a low yield coupled with a prohibitive cost of transportation.

The maps that the Lérys produced were based on material that had been submitted to them; neither mapmaker ever visited Lake Superior. The Léry's 1735 chart includes the two fanciful islands on a map for the first time, and assigns names to those islands and other extant islands in Lake Superior. The names appearing on the Léry chart are those of men who were related to La Ronde's mining project: "Phelippeaux" and "Pontchartrain" honor Jean-Frederic Phélypaux, Comte de Maurepas, who followed his father, the Comte de Pontchartrain, as Minister of the Marine, and assumed his father's title. Thus, one man's name appears on the three largest islands on Lake Superior, two of which do not exist.

The names of "Pontchartrain" and "Maurepas" were first placed on the North American continent at the end of the seventeenth century. In March 1699, Pierre le Moyne, Sieur d'Iberville, and his younger bother, Jean Baptiste le Moyne, Sieur de Bienville, went up the Mississippi River to Baton Rouge. On the return trip, they took a shortcut through the Bayou Manchac and came upon two lakes. D'Iberville named both for the powerful Phélypeaux family, which had sponsored their expedition. The larger Lake Pontchartrain honored Louis de Phélypeaux, Comte de Pontchartrain, who was the royal Minister of Marine at the time. The smaller Lake Maurepas took the name of Jérôme de Phélypaux, Comte de Maurepas, the Comte de Pontchartrain's son. When d'Iberville returned to France, the father resigned and the son became the royal Minister of Marine and, as such, the Comte de Pontchartrain. La Ronde had a direct association with d'Iberville and Bienville at the time when the two Louisiana lakes were named. In 1700, La Ronde was employed by d'Iberville to survey the mouth of the Mississippi River.

"I. Beauharnois" honored the Governor, whose name was also memorialized by having it attached to a river and harbor in the area. Beauharnois de Boishe was the cousin of Jean Frédéric Phélypeaux. The Intendant, Gilles Hocquart, was recognized by assigning his name to an island, a group of islets, and a point on the lake shore. The question has been raised as to whether "I. Ste. Anne" is a corruption of Saint-Ange, who underwrote the La Ronde mining expedition on Lake Superior.

All three charts by the Lérys bear the stamp of the Dépôt des cartes because they resided in the Dépôt des cartes et plans de la Marine in Paris. Therefore, they were readily available to Jacques-Nicolas Bellin, the chief engineer of the Dépôt. Bellin's most influential map, "Carte des Lacs du Canada" (see figure 61), was published in 1744 in the widely read *Histoire et description générale de la Nouvelle France* by Father Pierre-François-Xavier Charlevoix, who had been dispatched in 1720 by Louis XV, with Philippe, Duc d'Orléans, as regent, to determine if a large sea existed west of Lake Superior. The journey was authorized by Louis Alexandre de Bourbon, Comte de Toulouse, who was Director of the Council of the Marine. Charlevoix never visited Lake Superior. His travels led him across Lake Ontario, Lake Huron, Lake Michigan, and down the Mississippi to its mouth, whence he returned to France.

Bellin's seminal 1744 printed map depicts all five Great Lakes, and several islands are evident on Lake Superior. "Isle Royale" is the name on the largest island, which had been called I. Minong on previous maps. The second largest island, located just southeast of Isle Royale, is inscribed "I. Philippeaux au I. Minong." It and "I. Pontchartrain" constitute the two fictitious islands. "Isle Maurepas" is currently called

Michipicoton Island; "I. Hocquart" is Leach Island; "I. Beauharnois" is Montreal Island; "I. Ste Anne" is Caribou Island. The representation of Lake Ontario is thought to derive from a map by Léry, while the other Great Lakes are modifications of works by Franquelin and de l'Isle, with new place-names added..

In his "Remarques" on the maps that appeared in the preface to Charlevoix's *Journal historique d'un voyage fait par ordre du roi dans l'Amérique septentrionnale*, which accompanied his *Histoire et description générale de la Nouvelle France,* Bellin indicated that he used the manuscript material of the Dépôt for the outline of Lake Superior. Nowhere does Bellin mention the origin of the names he assigned to islands in the lake. In 1755, Bellin's "Partie Occidentale de la Nouvelle France ou Canada" was published separately as an update of the 1744 map. On that map, the outlines of Lake Ontario and Erie were altered, while the islands in Lake Superior persist.

Two of Bellin's French contemporary cartographers were Gilles Robert de Vaugondy and Jean-Baptiste Bourguignon D'Anville. Robert de Vaugondy published maps in his *Atlas portatif, universel et militaire* . . . , 1748–49, based on Bellin's representations of the Great Lakes. Similarly, Robert de Vaugondy's "Carte des Pays connus sous le nom de Canada," published separately in 1753, drew from Bellin's depictions. D'Anville's maps provided a model for the Great Lakes that differed somewhat from that shown by Bellin. The French maps, particularly those of D'Annville, were produced, in part, to assert French claims in North America. D'Anville's 1755 map "Carte Louisiane et Terres Angloises" not only makes territorial claims but even shows Lake Michigan with an incorrect southwesterly tilt, to place more land within control of the French. On Lake Superior, Isle Royal is called "I. Minong"; Bellin's "I. Philippeaux au I. Minong" is absent while the other islands, which were included on the Bellin map, are present and designated by the same names as appeared on the Bellin map.

The year 1755 witnessed several critical battles in the French and Indian War between Great Britain and France, which was not formally declared until May 18–19, 1756. Also in 1755, several English maps were published to rebut the French cartographic claims. The most famous of these maps was made by John Mitchell and published in London on February 13, 1755 (see figure 62). The Board of Trade retained Mitchell, a Virginia-born physician who had emigrated to England, to produce a map from information compiled by all the colonial governors. The first edition was entitled "A Map of the British and French Dominions in North America with the Roads, Distances, Limits, and Extent of the Settlements . . . ." The title was changed in the fourth edition of the map to "A Map of the British Colonies in North America . . . ." This frankly political map depicts the division of the eastern portion of North America between the British and the French. The boundaries of the English colonies extend westward across the Mississippi River to the western border of the map, reflecting their "Sea to Sea" charter.

The Mitchell map is regarded by most authorities as the most important map in American history. It is known that one of the later versions of the Mitchell map was consulted by the official representatives of both the United States and Great Britain at Paris in 1782 and 1783 in negotiating the treaty that terminated the Revolutionary War and established the boundaries of the United States of America. John Adams, Benjamin Franklin, and John Jay each has written regarding his use of that specific map. The copy of the Mitchell map that was used by Richard Oswald, representing the British King, resides in the British Library. Lawrence Martin, Chief of the Library of Congress Map Division from 1924 to 1946, wrote in the *Dictionary of American Biography*, "Without serious doubt Mitchell's is the most important map in American history."

The Mitchell map reproduces Bellin's representation of islands in Lake Superior, and even retains the names used on the French map. The two apocryphal islands, "I. Philippeaux" and "I. Pontchartain," are accompanied by "I. Royale," "I. Maurepas," "I. Hocquart," "I. Beauharnois," and "I. Ste. Anne" on the Mitchell map. Acceptance of the fictitious islands at the time of the 1783 Treaty of Paris is attested to by Article II of that treaty, which states that the boundary between the United States and British North America was to run "through Lake Superior northward of the Isles Royal and Phelipeaux . . . ." Thus, the original geographic limits of United States were based on an island that did not exist.

Other English maps were published in 1755 as part of a campaign to assert British claims and counter French claims to land in North America. John Huske's map, "A New and Accurate Map of North America," published in *The Present State of North America, &c. Part I*, includes "Pontchartrain," "Maurepas," "Hocquart," and "I^{s} S^{te}. Ann." Isle Royale is called "Minong" and Phillipeaux is absent. On "A New and Accurate Map of the English Empire in North America . . . By a Society of Anti-Gallicans," "I. Royale," "I. Philippeaux," "I. Pontchartrain," "I. Maurepas," and "I. S^{t}. Ann" are shown, while Hocquart and Beauharnois do not appear.

Conrad E. Heidenreich reported that, in July 1823, Major Joseph Delafield, agent to the American Boundary Commission responsible for surveying the international boundary through Lake Superior, noted in his journal that "nobody seems to know any islands by the names of Phillipeaux . . . ." The last map in the National Map Collection, Ottawa, to include a fictitious island in Lake Superior is Gall and Inglis's "Map of the United States," published in Edinburgh in 1842. On that map, Isle Philippeaux is located off the Keweenaw Peninsula of the south shore of the lake.

Thus, one of the two fanciful islands that were inserted on Lake Superior persisted on maps for over a century. Both fanciful islands were accepted by the contemporary British cartographers, and "I. Phelypeaux" avoided deletion long enough to be cited as a specific reference point in the treaty that established the boundary of the United States of America. The inappropriate presence of the island on a map resulted in its inclusion in the definition of a new nation.

References

Cartier, Jacques. *Bref Récit et succincte narration de la navigation faite in MDXXXV et MDXXXVI*. Paris: Tross, 1863.

Champlain, Samuel. *Des Sauvages, ou, Voyages de Samuel Champlain, de Brouage, fait en la France nouuelle, l'an mil six cens trois . . . .* Paris, 1603.

———. *Les Voyages*. Paris, 1613.

———. *Les Voyages de la Novvelle France occidentale, dicte Canada, faits par le Sr de Champlain*. Paris, 1632.

———. *Voyages et descouvertes faites en la Nouvelle France depuis l'année 1615 iusques à la fin de l'année 1618*. Paris, 1619.

Charlevoix, Father Pierre-François-Xavier. *Histoire et description générale de la Nouvelle France*. Paris, 1744.

———. *Journal historique d'un voyage fait par ordre du roi dans l'Amérique septentrionnale*. Paris, 1744.

Chatelain, Henri Abraham. *Atlas historique*. 7 vols. Amsterdam, 1732–39.

De l'Isle, Joseph Nicholas. *Nouvelle Cartes des découvertes de l'Amiral de Fonte*. Paris, 1753.

Delafield, Joseph. *The Undefined Boundary: A Diary of the First Survey of the Canadian Boundary Line St. Regis to Lake of the Woods*. Edited by R. McElroy and T. Riggs. New York: Private printing, 1939.

Dobbs, Arthur. *An Account of the Countries Adjoining to Hudson's Bay in the north-west part of America*. London, 1744.

Faillon, Etienne. *Histoire de la colonie française en Canada*. 3 vols. Montreal: Bibliothèque paroissale, 1865–66.

Fer, Nicolas de. *L'Atlas curieux*. Paris, 1700–1703.

Hennepin, Louis. *Nouvelle Découverte d'un très grand pays situé dans l'Amérique, entre le Nouveau Mexique et la Mer glaciale*. Utrecht, 1697.

Jesuit Relations. 1610–1791.
Lahontan, Louis-Armand de Lom d'Arc, Baron de. *Nouveaux Voyages de Mr le baron de Lahontan dans l'Amérique Septentrionale*. The Hague, 1703. English edition, London, 1703.
Parkman, Francis. *La Salle and the Discovery of the Great West*. Boston: Little, Brown, and Co., 1879.
Robert de Vaugondy, Gilles. *Atlas portatif, universel et militaire*. Paris, 1748–49.
Sagard Théodat, Gabriel. *Le Grand Voyage du pays des Hurons*. Paris, 1632.
Wytfliet, Cornelis van. *Descriptionis Ptolemaicae Augmentum*. Louvain, 1597.

SUGGESTED READING

Cumming, W. P., S. E. Hillier, D. B. Quinn, and G. William. *The Exploration of North America, 1630–1776*. New York: G. P. Putnam's Sons, 1974.
Delanglez, Jean. *Life and Voyages of Louis Jolliet (1645–1700)*. Chicago: Institute of Jesuit History, 1948.
Gaither, Frances. *The Fatal River: The Life and Death of La Salle*. New York: Henry Holt, 1931.
Hatcher, Harlan. *The Great Lakes*. London, New York: Oxford University Press, 1944.
Heidenreich, Conrad E. "Explorations and Mapping of Samuel de Champlain, 1603–1632." *Cartographica* Monograph No. 17 (1976).
———. "The Fictitious Islands of Lake Superior." *Inland Seas* 43, no. 3 (1987): 168–77.
———. "Mapping the Great Lakes: The Period of Exploration, 1603–1700." *Cartographica* 17, no. 3 (1980): 32–64.
———. "Mapping the Great Lakes: The Period of Imperial Rivalries, 1700–1760." *Cartographica* 18, no. 3 (1981): 74–109.
———. "Seventeenth-Century Maps of the Great Lakes: An Overview and Procedures for Analysis." *Archivaria* 6 (1978): 83–112.
Karrow, Robert W. Jr. "Lake Superior's Mythic Isles: A Cautionary Tale for Users of Old Maps." *Michigan History* 69, no. 1 (1985): 24–31.
Magnaghi, Russell M. "The Jesuits in Lake Superior Country." *Inland Seas* 41, no. 3 (1985): 190–203.
Morison, Samuel Eliot. *Samuel de Champlain, Father of New France*. Boston: Little, Brown and Company, 1972.
Ristow, Walter W. "John Mitchell's Map of the British and French Dominions in North America," pp. 102–13. In *A la Carte*. Washington, D.C.: Library of Congress, 1972.
Roberts, W. Adolphe. *Lake Pontchartrain*. Indianapolis: The Bobbs-Merrill Company, 1946.
Schwartz, Seymour I. *The French and Indian War, 1754–1763: The Imperial Struggle for North America*. New York: Simon & Schuster, 1994.
Schwartz, Seymour I., and Ralph E. Ehrenberg. *The Mapping of America*. New York: Harry N. Abrams, 1980.

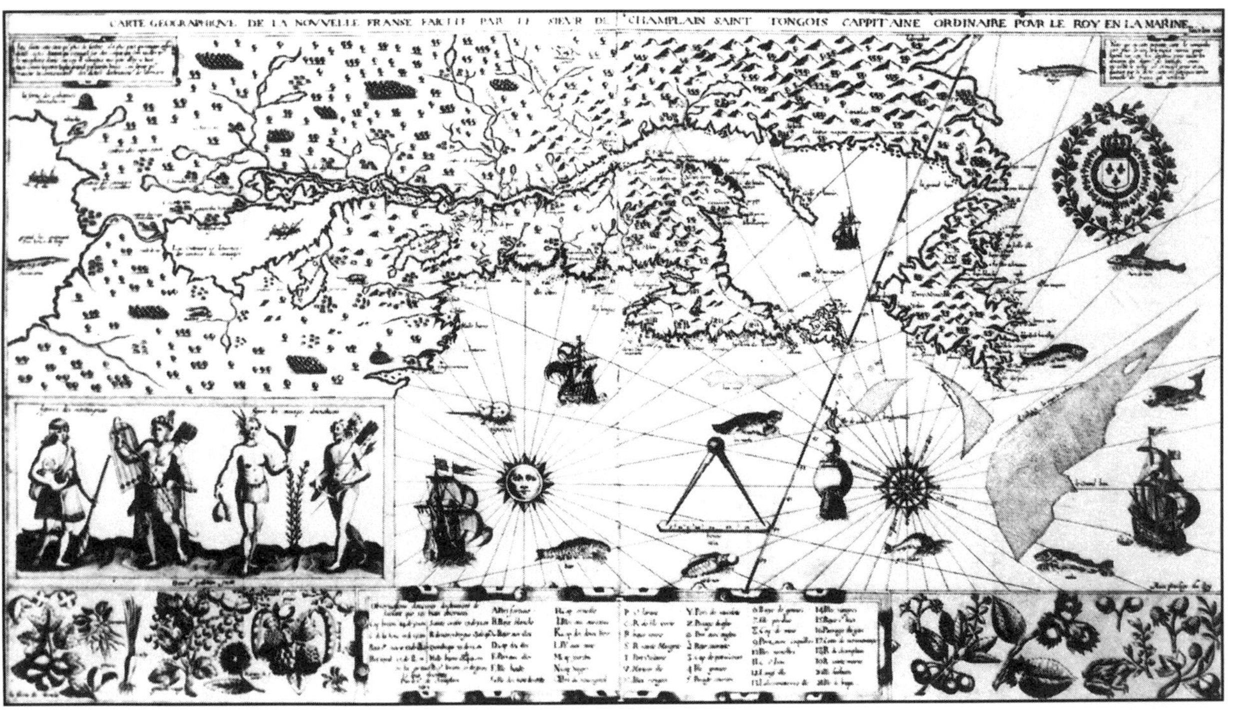

Fig. 51. Samuel de Champlain. "CARTE GEOGRAPHIQVE DE LA NOVVELLE FRANSE FAICTTE PAR LE SIEVR DE CHAMPLAIN SAINT TONGOIS CAPPITAINE ORDINAIRE POVR LE ROY EN LA MARINE." Copperplate, 440 x 765 mm. From *Les Voyages du Sieur de Champlain* (Paris 1613). Private Collection.

The first map to indicate a chain of Great Lakes. The map was drawn before Champlain had personally seen Lake Ontario and Lake Huron. It was based on the descriptions of Native Americans.

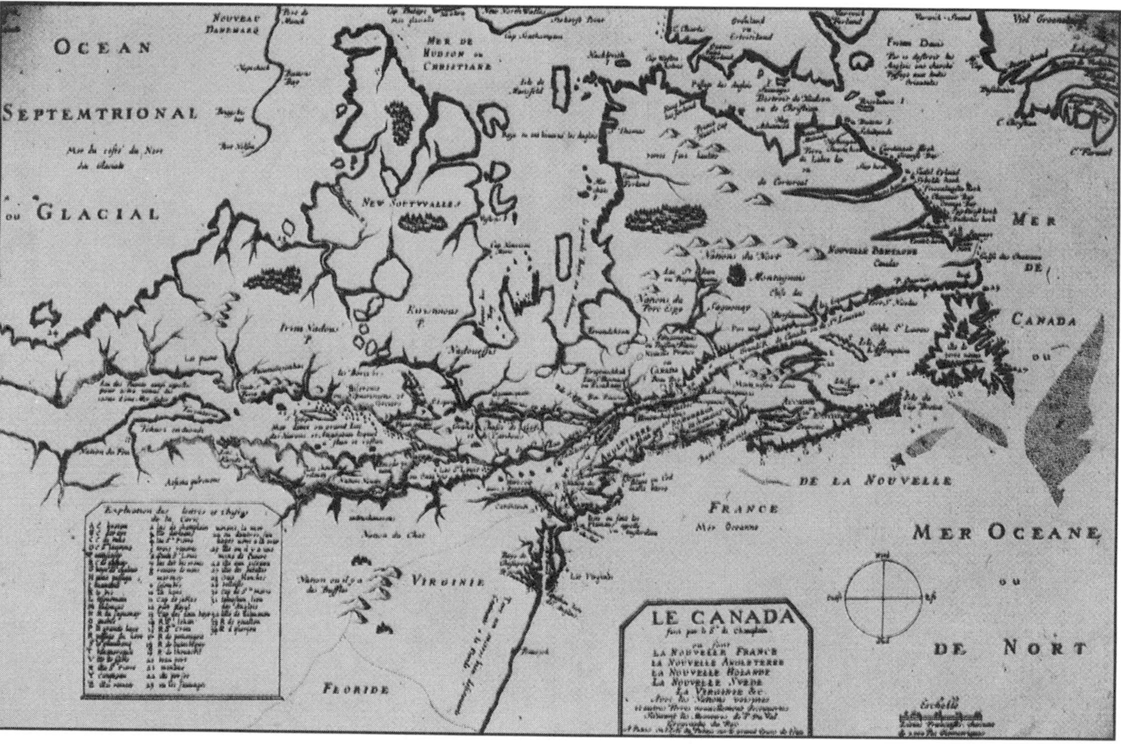

Fig. 52. Pierre Duval. "LE CANADA faict par le Sr Champlain, où sont LA NOUVELLE FRANCE, LA NOUVELLE ANGLETERRE, LA NOUVELLE HOLANDE, LA NOUVELLE SVEDE, LA VIRGINIE &C . . . ." Paris, 1653. Copperplate, 345 x 540 mm. Private Collection.

The map used a copperplate made in 1616, which produced the first map to depict part of the Great Lakes based on European exploration.

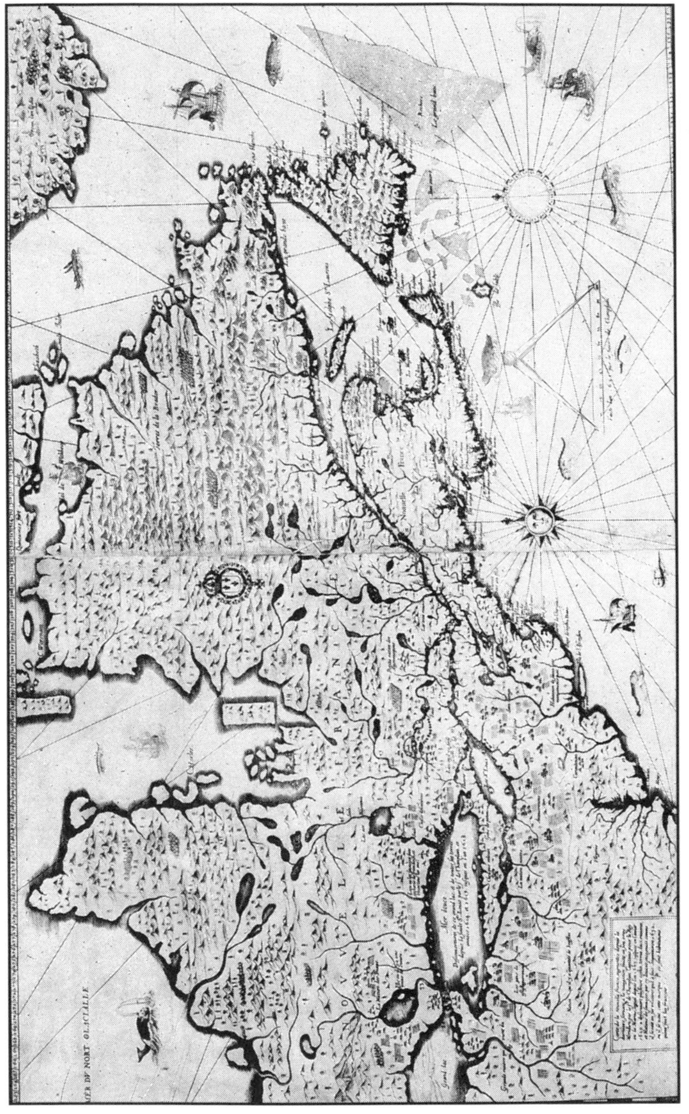

Fig. 53. Samuel de Champlain. "Carte de nouuelle France, augmenté depuis la derniere seruant a la navigation faicte en son vray Meridien par le Sr de Champlain Capitaine pour le Roy en la Marine . . . . Faicte l'an 1632 par le sieur de Champlain." Copperplate, 525 x 870 mm. From *Les Voyages de la Nouvelle France occidentale, dicte Canada, faits par le Sr de Champlain* (Paris, 1632). Private Collection.

The map depicts significant changes in the representation of "Lac St. Louis" (Lake Ontario) and "Mer Douce" (Lakes Huron and Michigan). "Grand lac" is also shown.

Fig. 54. Nicolas Sanson. "AMERIQUE SEPTENTRIONALE Par N. Sanson . . . ." Copperplate, 390 x 555 mm. Paris, 1650. Private Collection

The first printed map to delineate the five Great Lakes in a recognizable form. The first map to name Lakes Ontario and Superior.

Fig. 55a. Francesco Guiseppe Bressani. 'Novæ Franciæ Delineato 1657." Copperplate, 510 x 745 mm. Macerata?, 1657. Photograph courtesy of National Map Collection, National Archives of Canada, Ottawa, NMC-6338, NMC-194824. The left half of this map depicts the Great Lakes.

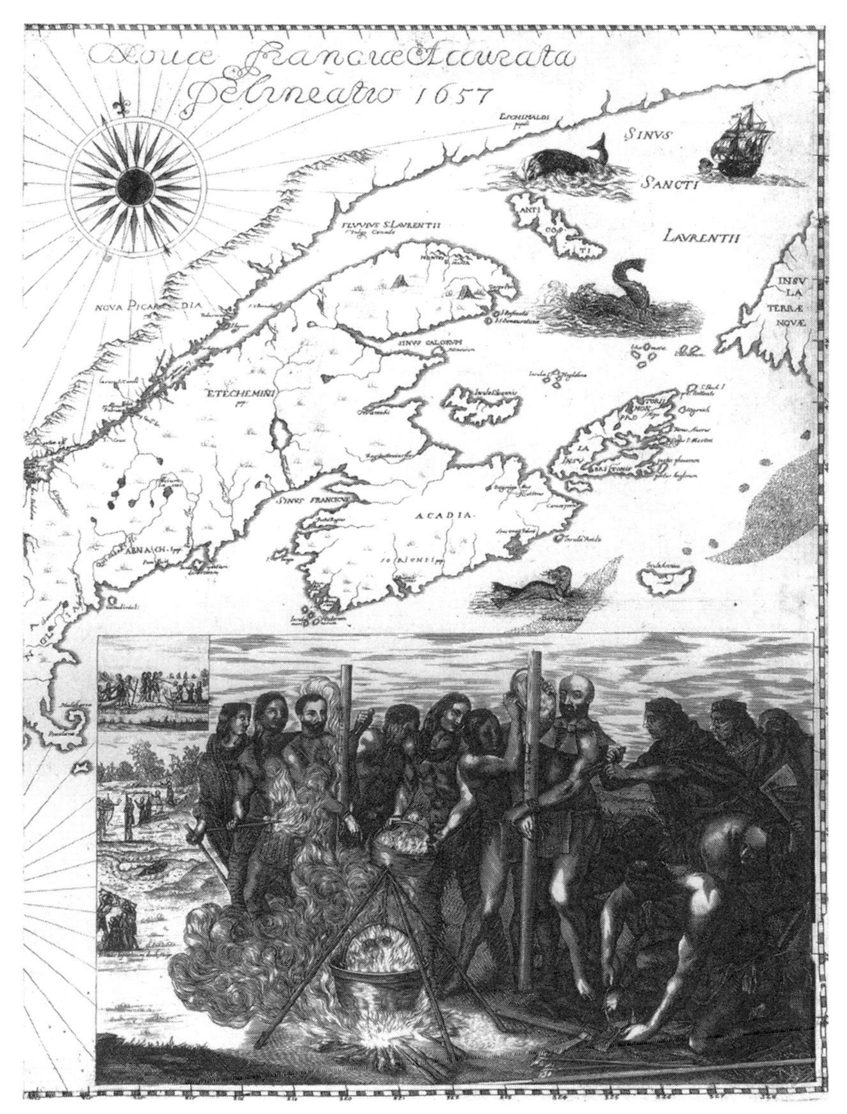

Fig. 55b. This right half depicts the St. Lawrence Region and the martyrdom of Fathers Brébant and Lalemant.

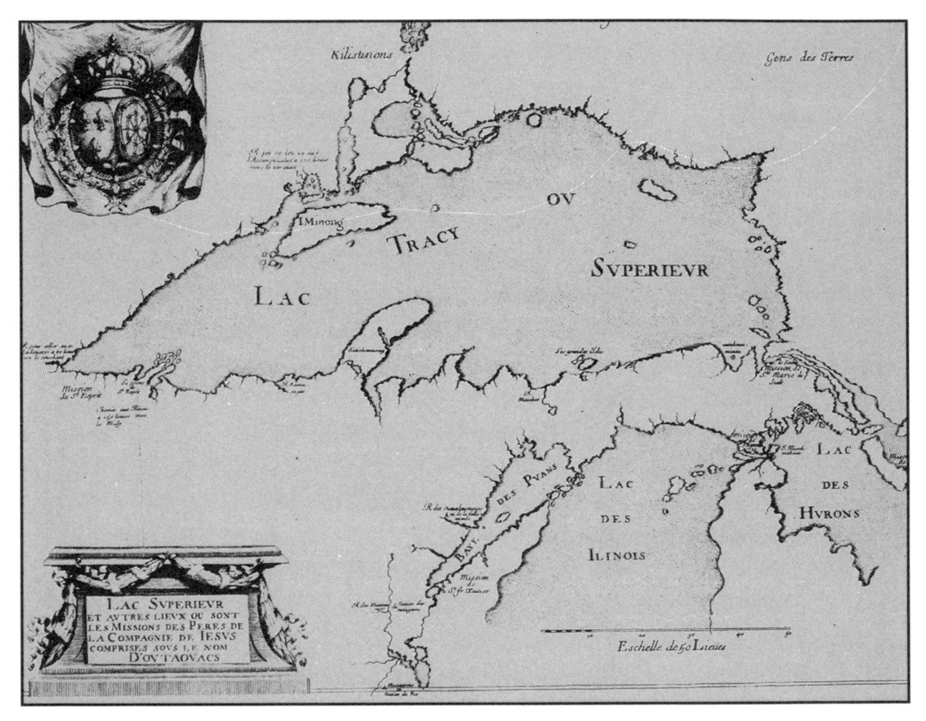

Fig. 56. Father Claude Dablon: "LAC SVPERIEVR ET AVTRES LIEVX OU SONT LES MISSIONS DES PERES DE LA COMPAIGNE DE JESVS COMPRISES SOVS LE NOM D'OVTAOVACS." Copperplate, 350 x 480 mm. From Dablon, *Relation de ce que s'est passé de plus remarquable . . . en la Nouvelle France . . . 1671* (Paris, 1672). Private Collection.

The map is the earliest printed survey of Lake Superior. It is based on the explorations of C. Allouez, S. J.

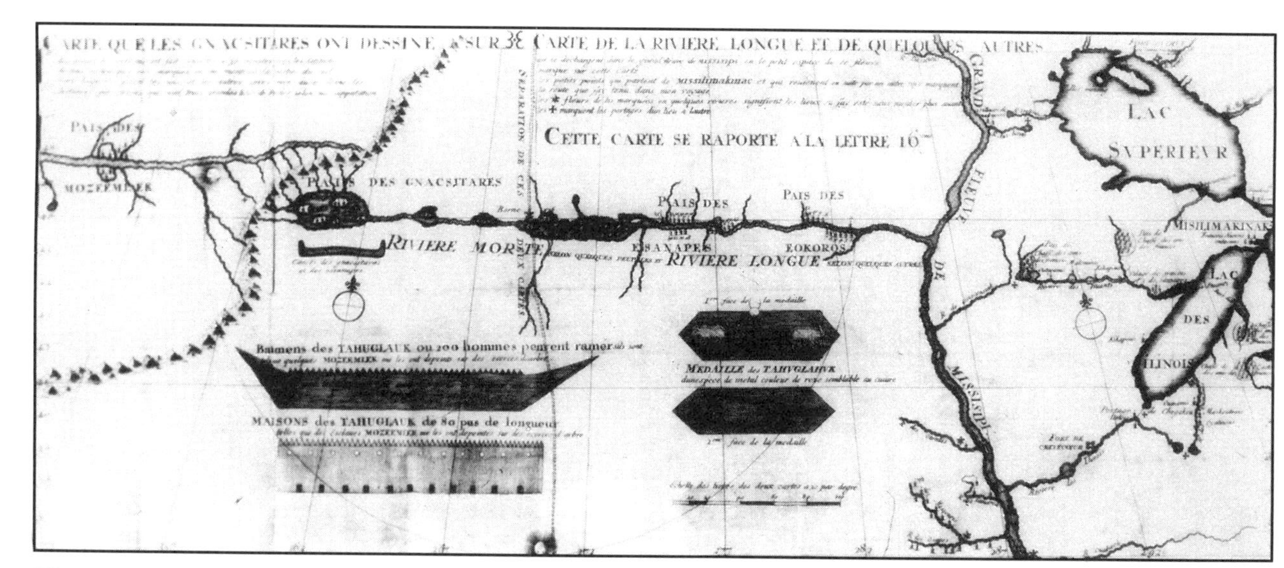

Fig. 57. Louis Armand de Lom d'Arce, Baron de Lahontan. "CARTE QUE LES GNACSITARES ONT DESSINÉ SUR CARTE DE LA RIVIERE LONGUE ET DE QUELQUES AUTRES." Copperplate, 280 x 650 mm. From Lahontan, *Nouveaux Voyages de Mr. Baron de Lahontan dans l'Amérique septentrionale* (The Hague, 1703). Private Collection.

The map appears in a book that underwent twenty-five editions or versions. The map depicts an apocryphal "Long River," which flowed into a large lake to the west, suggesting a waterway to the Pacific Ocean.

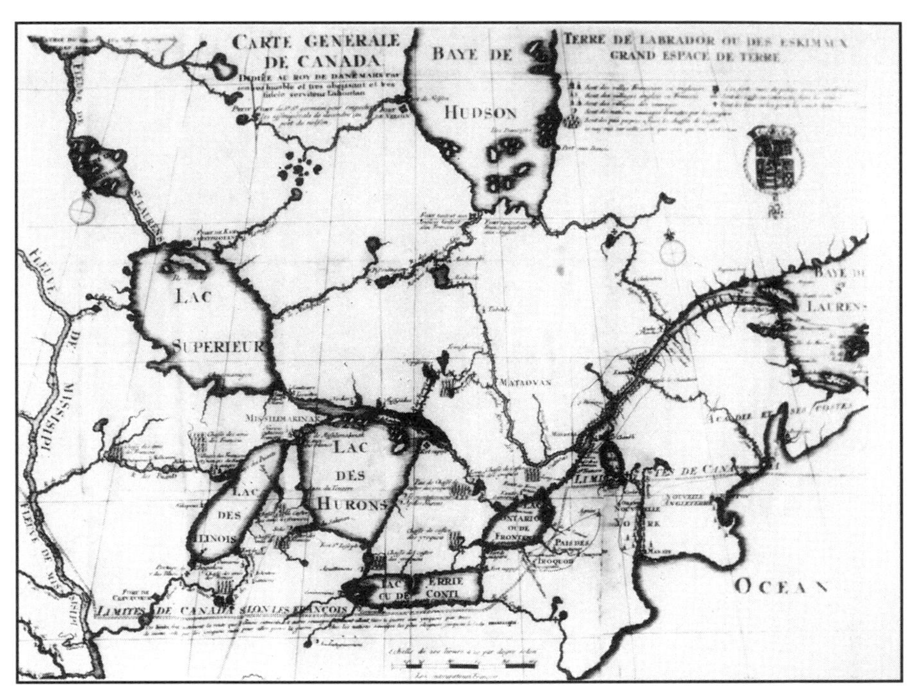

Fig. 58. Louis Armand de Lom d'Arce, Baron de Lahontan. "CARTE GENERALE DE CANADA." Copperplate, 380 x 530 mm. From Lahontan, *Nouveaux Voyages de Mr. Baron de Lahontan dans L'Amérique septentrionale* (The Hague, 1703). Private Collection.

The map shows an inaccurate geography of the Great Lakes.

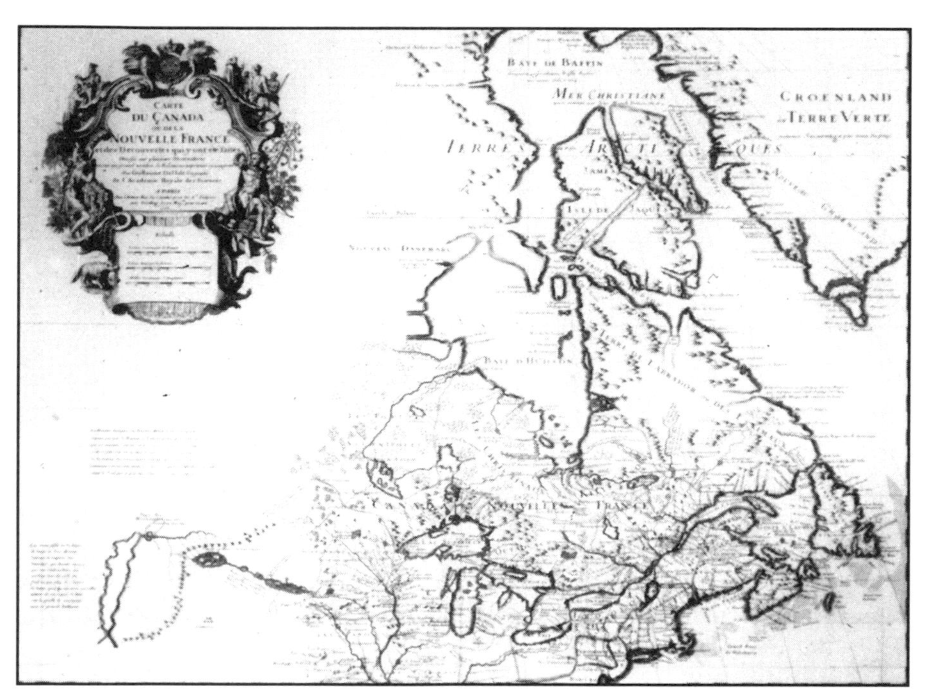

Fig. 59. Guillaume de l'Isle. "CARTE DU CANADA OU DE LA NOUVELLE FRANCE . . . ." Copperplate, 510 x 660 mm. Paris, 1703. Private Collection.

The Great Lakes are shown correctly. Only one island, "I. Minong," is named in Lake Superior.

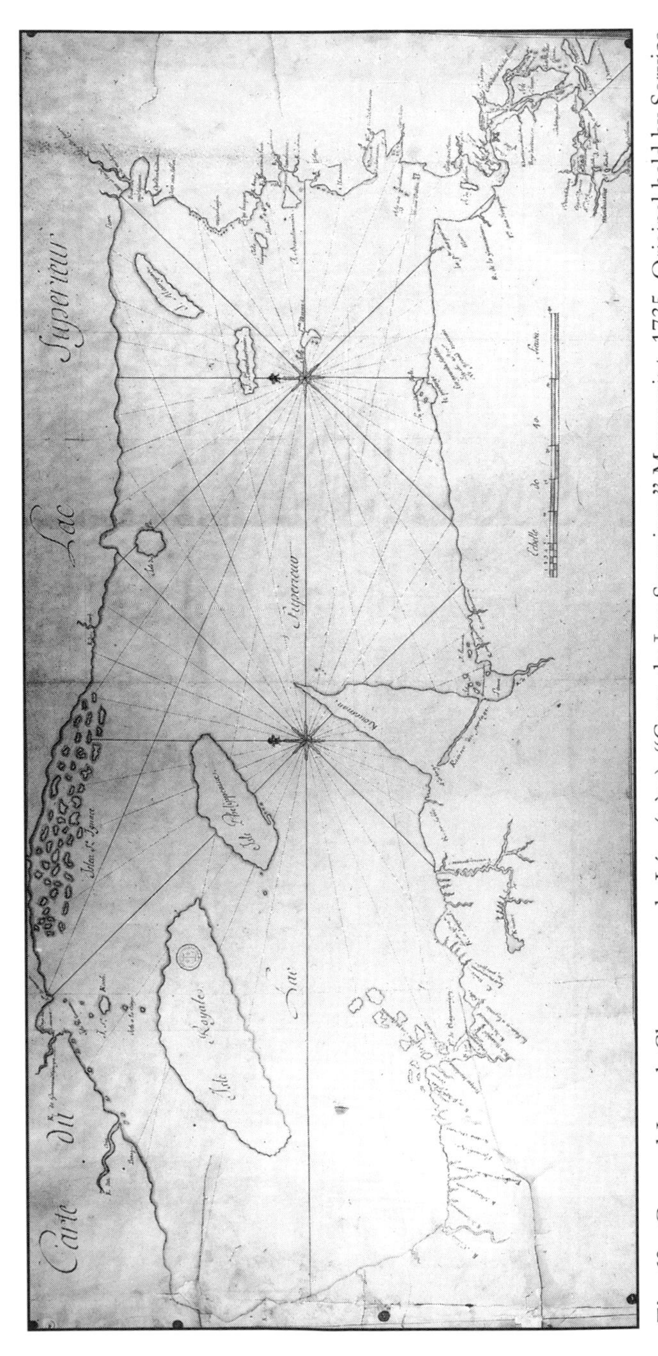

Fig. 60. Gaspard-Joseph Chaussegros de Léry (père). "Carte du Lac Superieur." Manuscript, 1735. Original held by Service historique de la Marine, Vincennes. From a photographic copy held by National Archives of Canada, Ottawa; reproduced with permission of the Service historique de la Marine and the National Archives of Canada, NMC-29015.

This manuscript shows the two fictitious islands, "Isle Phelippeaux" and "I. Pontchartrain."

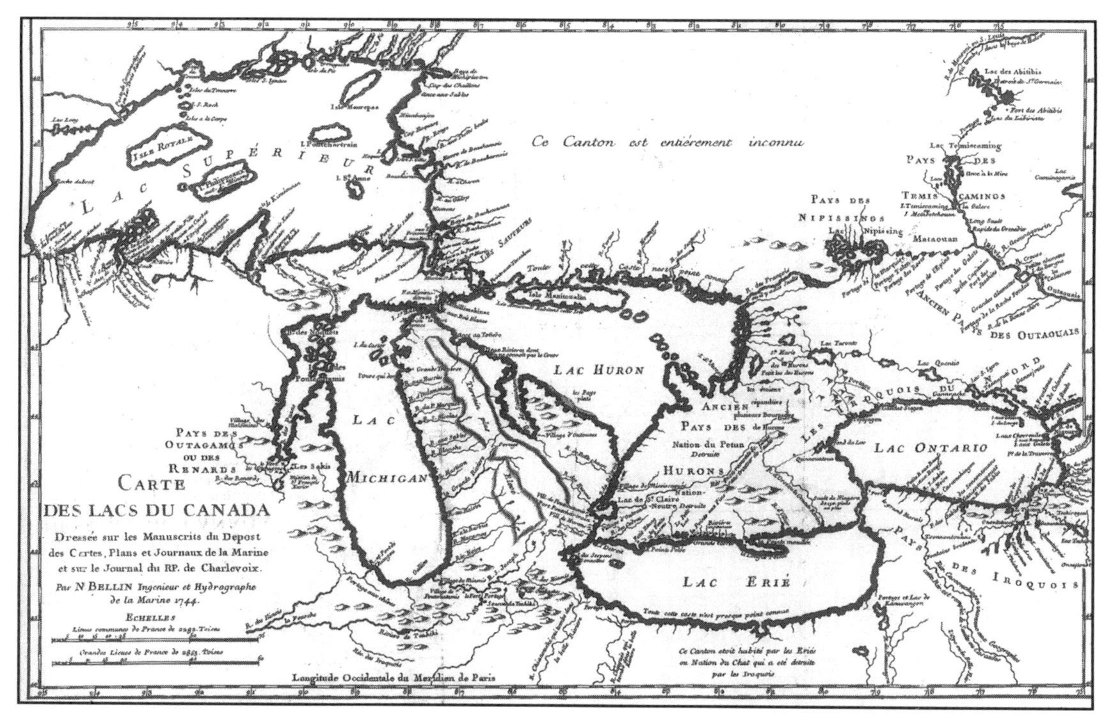

Fig. 61. Jacques-Nicolas Bellin. "CARTES DES LACS DU CANADA" Copperplate, 270 x 490 mm. From Father Pierre-François-Xavier Charlevoix, *Histoire et description générale de la Nouvelle France* (Paris, 1744). Photograph courtesy of National Archives of Canada, Ottawa, NMC-6425.

The first printed map to incorporate information from the Léry's manuscripts and include the fanciful islands in Lake Superior .

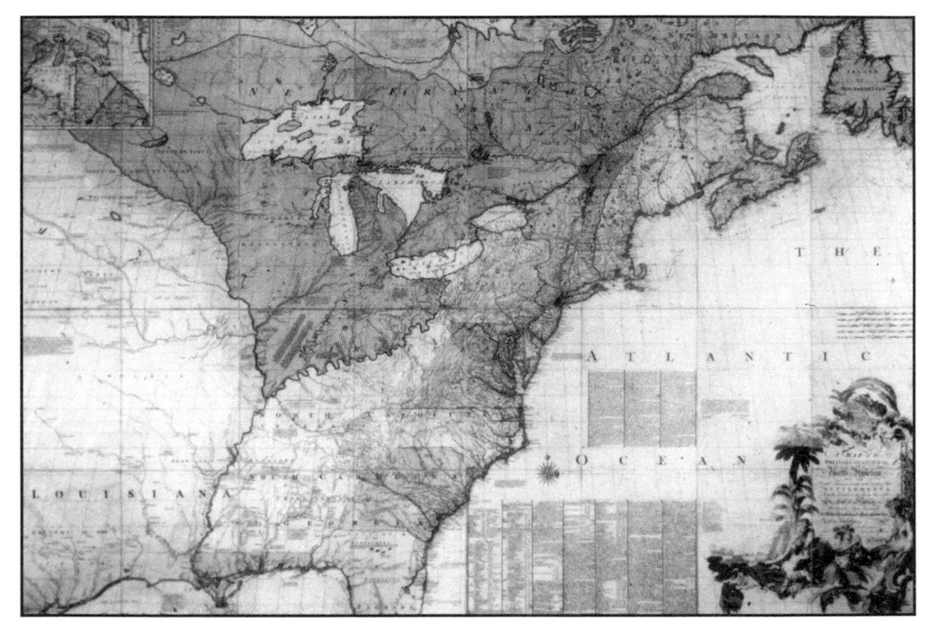

Fig. 62. John Mitchell. "A Map of the British and French Dominions in North America with the Roads, Distances, Limits and Extent of the Settlements. . . ." London, 1755. Copperplate, eight sheets, each 980 x 760 mm. Private Collection.

This is the map that was used by the three negotiators from the United States and the representative of King George III in establishing boundaries at the Treaty of Paris in 1783.

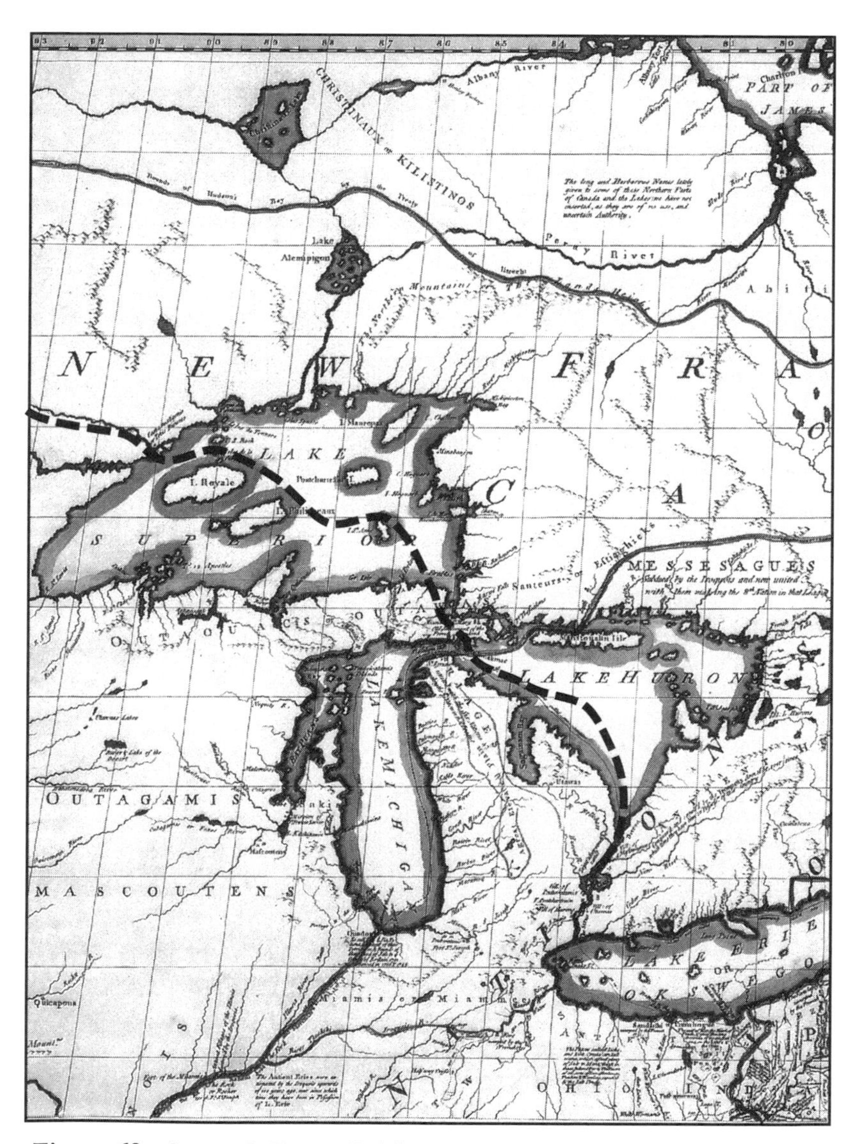

Figure 62a. Inset of Figure 62. The sheet, one of eight, that includes the apocryphal "I. Philippeaux" and "I. Pontchartrain." The boundary line has been superimposed to indicate its relation to I. Philippeaux.

CONCLUSION

Misrepresentations of geography have appeared in narratives and on maps throughout history. Tales of travels, particularly to unfamiliar lands, lend themselves to embellishment. Either a vivid imagination or a deliberate deception, which may be difficult to refute, often adds appeal for the audience. A dissertation on geography can be dull, but, if romantic infusions are included to make the land more attractive or potentially more valuable, the subject matter will excite interest.

During the twelfth century, long before the New World was discovered, a frankly false but appealing and influential tale appeared. It was reported that, somewhere in Asia, there reigned a wealthy Christian monarch, named Presbyter John or Prestor John. He was thought to be a direct descendant of the Magi. Many years ago, he was supposedly willing to join forces with the western Christians in their battles against the evil, albeit mythical, Gogs and Magogs. The quest for Prestor John's intriguing land of unmatched wealth was a stimulus for exploration of Asia. The Empire of Prestor John was the subject of many medieval maps, even a sixteenth-century map by the greatest cartographer of his time, Abraham Ortelius. Despite its appearance in Asia or Africa on printed maps, Prestor John's land of riches was an unequivocal fabrication. In the fourteenth century, Sir John Mandeville's Travels and Voyages, describing wondrous sights that he had seen during his alleged travels afar, was one of the most widely read medieval books. Mandeville, in fact, probably never traveled from his home and was a believable but blatant liar.

From our current vantage point, we appreciate that the expansion of geographic knowledge has been accompanied by an increased precision is defining and disseminating that knowledge. As new lands were discovered, so new techniques were applied to define the discoveries. During the age of discovery, the geography was determined by estimates made by the naked eye. And later, the determinations were refined and made more exact by telescopic viewing and the application of astronomic tables. In more modern times, aerial photography and, subsequently, satellite mapping removed speculation regarding the geography. Most recently, the Geographic Positioning System has added to the accuracy.

Interposed chronologically between the myths of Prestor John and the modern era of scientific sophistication was a time of the discovery and early settlement of what would become the United States of America. It was a time that preceded the availability of tools to affirm the truth, and sightings were occasionally accompanied by speculations.

The discovery of the "New World" provided a new and exciting focus for geographic description and cartographic depiction. Within a decade of the discovery of a continental mass in the southern portion of the Western Hemisphere, a name was assigned to and figuratively inscribed on that land. An interpretation of perhaps only two vague printed narratives brought about the permanent placement of the greatest misnomer on Planet Earth. Fiction was accepted as fact. Although historians in the past have affirmed Christopher Columbus's primacy of discovery, his name will never designate the continent he discovered.

As the attention of the European powers in the sixteenth century incorporated consideration of the continental land in the northen part of the Western Hemisphere, geographic facts and fantasies continued to be juxtaposed in text and on charts. Those who chronicled and mapped the exploits of explorers of the newly discovered land generally resided in scriptoria, publishing houses, and mapmaking venues far afield from the sites of exploration. The opportunity for individual misinterpretation by both the explorers and the chroniclers was rampant, and the initial misinterpretations were compounded with time. Interposed between new exciting and informative facts in narratives and on maps, for the ensuing three centuries, there appeared misinformation that was often equally or more exciting.

History is molded and, in part, defined by the historian. Just as the interpretation of history is subject to change, so concepts of the geography of the land have not been immutable. Pliny the Elder in his first century A.D. discussion of botanists wrote that "Pictures are apt to mislead." Maps, as pictures, surely have misled and deceived. Maps of America were not immune to the inclusion of errors and apocrypha. As indicated by Robert W. Karrow, Jr., this was recognized in the conversation between Huck Finn and Tom Sawyer in Mark Twain's *Tom Sawyer Abroad:*

—Tom Sawyer, what's a map for? Ain't it to learn you facts?
—Of course.
—Well then, how's it going to do that if it tells lies? That's what I want to know.

Our current age of technology is not a time to minimize or belittle the accomplishments of the past. In most instances, interpretations and reports were correct. Major advances were made with a minimum of tools. As both historians and cartographers have demonstrated, although the past may be permanent, presentation of the past is subject to change. The errors in reporting that have occurred should be looked upon in the context of the time. The facts presented on maps are not absolute truths. Just as the land itself is subject to tectonic shifts, so the depictions of that land might change. The Mismapping of America is but a small part of the Mapping of America, but that small part has had fascinating and significant consequences.

Index

A

B

C

D

E

F

G

H

M

N

O

P

Q

R

S

T

U

V

W

Y

Z